Construction
Management
Strategies

The book's companion website at www.wiley.com/go/constructionmanagementstrategies offers invaluable resources for students and lecturers as well as for practising construction managers:

- end-of-chapter exercises + outline answers
- Gantt charts to accompany examples in the book
- PowerPoint slides for each chapter
- ideas for discussion topics
- links to useful websites

Construction Management Strategies

A Theory of Construction Management

Milan Radosavljevic
Lecturer, Course Director of the
MSc in Project Management,
The University of Reading
UK

John Bennett
Professor Emeritus
The University of Reading
UK

A John Wiley & Sons, Ltd., Publication

This edition first published 2012
© 2012 John Wiley & Sons, Ltd

Blackwell Publishing was acquired by John Wiley & Sons in February 2007.
Blackwell's publishing program has been merged with Wiley's global
Scientific, Technical and Medical business to form Wiley-Blackwell.

Registered office:
John Wiley & Sons, Ltd, The Atrium, Southern Gate, Chichester, West
Sussex, PO19 8SQ, UK

Editorial offices:
9600 Garsington Road, Oxford, OX4 2DQ, UK
The Atrium, Southern Gate, Chichester, West Sussex, PO19 8SQ, UK
2121 State Avenue, Ames, Iowa 50014-8300, USA

For details of our global editorial offices, for customer services and for information
about how to apply for permission to reuse the copyright material in this book
please see our website at www.wiley.com/wiley-blackwell.

Library of Congress Cataloging-in-Publication Data

Radosavljevic, Milan.
   Construction management strategies : a theory of construction management /
Milan Radosavljevic, John Bennett.
       p. cm.
   Includes bibliographical references and index.
   ISBN 978-0-470-65609-9 (pbk. : alk. paper)
  1. Construction industry–Management–Study and teaching.   I. Bennett, John,
1936-   II. Title.
   TH438.R324 2012
   624.068'4–dc23

                                                                2011035228

A catalogue record for this book is available from the British Library.

Wiley also publishes its books in a variety of electronic formats. Some content that
appears in print may not be available in electronic books.

Set in by 9/11.5pt Avenir-Roman by Thomson Digital, India

1   2012

Contents

The book's companion website at
www.wiley.com/go/constructionmanagementstrategies
offers invaluable resources for students and lecturers as well
as for practising construction managers:
- end-of-chapter exercises + outline answers
- Gantt charts to accompany examples in the book
- PowerPoint slides for each chapter
- ideas for discussion topics
- links to useful websites

With an ambition to multiply its customers' potential to think and achieve big, Tekla provides a BIM (Building Information Modeling) software environment that can be shared by contractors, structural engineers, steel detailers and fabricators, as well as concrete detailers and manufacturers.

The highly detailed as-built 3D models created, combined and distributed with Tekla software enable the highest level of constructability and production control. Centralizing building information into the model allows for more collaborative and integrated project management and delivery. This translates into increased productivity and elimination of waste, thus making construction and buildings more sustainable.

www.tekla.com

About the Authors

Dr Milan Radosavljevic, u.d.i.g. PhD ICIOB

Dr Milan Radosavljevic is a lecturer in Construction Management, Director of Postgraduate Programmes and Director of the prestigious MSc Project Management in the School of Construction Management and Engineering of the University of Reading.

Dr Radosavljevic currently serves as a committee member of the Association of Researchers in Construction Management (ARCOM), he sits on the Board of the International Council for Research and Innovation in Building and Construction (CIB), and represents the University of Reading in the UK Construction Industry Research and Information Association (CIRIA) and Constructing Excellence in the Built Environment (CEBE). He is also a Visiting Professor at the University of Ljubljana in Slovenia.

Before joining the University of Reading Dr Radosavljevic worked as a Production Director in the medium size off-site construction company in Central Europe from 1997 to 2001, as a Research Assistant at the Construction Information Technology Centre (CITC) of the University of Maribor in Slovenia where he administered the EU funded ITC Euromaster programme between 2001 and 2002, and as a Demonstration Projects Coordinator for Scotland on behalf of the Communities Scotland and Constructing Excellence between 2003 and 2006.

He has made contributions on various courses at institutions around the world, including the Norwegian School of Management, Tallinn University of Technology in Estonia, University of Ljubljana in Slovenia, Shandong University in China, and has developed and run executive courses in Project Management in the United Kingdom and abroad.

He was a Principal Investigator in the KanBIM project. This was an international project involving researchers from the University of Reading and Technion in Israel aimed at developing a Building Information Modelling (BIM) based lean production management system for construction jointly funded by the Innovative Construction Research Centre (ICRC) of the University of Reading and Tekla Oy, a major BIM software vendor from Finland. The initial year-long project culminated in a paper published by *Automation in Construction* journal where it has soon become the second most popular and downloaded research paper.

Apart from BIM and digital technologies in their broadest sense, his current research interests include programme and project management, and computational simulation of construction organizations as heterogeneous and evolving networks.

Key publications

Cus Babic, N., Rebolj, D., Magdic, A., and Radosavljevic, M (2003) MC as a means for supporting information flow in construction processes. *Concurrent Engineering: Research and Applications*, 11(1), 37–46.

Radosavljevic, M. (2008) Autopoiesis vs. social autopoiesis: critical evaluation and implications for understanding firms as autopoietic social systems. *International Journal of General Systems*, 37(2), 215–30.

Radosavljevic, M. and Horner, R.M.W (2002) The evidence of complex variability in construction labour productivity. *Construction Management and Economics*, 20(1), 3–12.

Radosavljevic, M. and Horner, R.M.W (2007) Process planning methodology: dynamic short-term planning for off-site construction in Slovenia. *Construction Management and Economics*, 25(2), 143–56.

Sacks, R., Radosavljevic, M. and Barak, R. (2010) Requirements for building information modeling based lean production management systems for construction. *Automation in Construction*, 19(5), 641–55.

Sergeeva, N. and Radosavljevic, M. (2011) Towards a Theoretical Framework for Creative Participation: How Personal Characteristics Influence Employees' Willingness to Contribute Ideas. In A. Mesquita, (ed.) *Technology for Creativity and Innovation: Tools, Techniques and Applications*. IGI Global.

Professor John Bennett, DSc, FRICS

John Bennett is Professor Emeritus of The University of Reading where he was Professor in the Department of Construction Management & Engineering from 1975 to 2001. He was the United Kingdom's first Professor of Quantity Surveying following a successful career in both the public and private sectors. This included being Senior Quantity Surveyor in the CLASP Development Group which pioneered the use of industrialised building and subsequently Chief Quantity Surveyor at Hampshire County Council.

Professor Bennett was Director of the Centre for Strategic Studies in Construction from 1986 to 1997 where he took the lead in publishing reports based on rigorous academic research that influenced practice. Important examples include *Building Britain 2001* and *Investing in Building 2001* which provide an action plan for UK construction endorsed by the then Prime Minister, Margaret Thatcher.

Professor Bennett's research provided the basis for the UK construction industry's approach to partnering. This is reflected in his influential publications: *Trusting the Team*, *The Seven Pillars of Partnering* and most recently *Partnering in the Construction Industry; Code of Practice for Strategic Collaborative Working*.

In 1991 Professor Bennett was employed as Professor in the Research Centre for Advanced Science and Technology (RCAST) at the University of Tokyo where he continued research, begun in 1985, into the management methods of the 'big five' Japanese contractors. In 2002 he was an international visitor providing strategic advice to the Australian Cooperative Research Centre for Construction Innovation based at the Queensland University of Technology in Brisbane.

He was the principle academic member of the consortium, led by W S Atkins International, which produced the Strategic Study on the Construction Sectors for the Commission of the European Union. He was one of two main authors of the Final Report, *Strategies for the European Construction Sector*, published in 1994 to provide a factual and theoretical basis for the EU's strategy towards construction.

He was Chairman of the SMM Development Unit that drafted SMM7 and the first Chairman of the joint ACE/BEC/RIBA/RICS Building Project Information Committee set up to run the UK's co-ordinated conventions for production information for building projects.

He was founding editor of the leading international refereed journal *Construction Management and Economics*, and remained in this role from 1982–91. His main theoretical publications are *International Construction Project Management*, which provides a contingency theory of construction project management based on practice in the United States, Japan and the United Kingdom and *Construction – The Third Way*, which describes construction organizations in terms of self organizing networks of teams guided by open information and feedback.

Publications

Atkins, W. S. and consultants including the Centre for Strategic Studies in Construction (1993) *Secteur, Strategic Study on the Construction Sector: Final report: strategies for the construction sector*. W.S. Atkins International.

Bennett, J. (1991) *International Construction Project Management: General theory and practice*. Butterworth Heinemann.

Bennett, J. (2000) *Construction – The Third Way: Managing Cooperation and Competition in Construction*. Butterworth.

Bennett, J., Croome, D. and Atkin, B. (1989) *Investing in Building 2001*. Centre for Strategic Studies in Construction.

Bennett, J., Flanagan, R., Lansley, P.R., Gray, C., Atkin, B.L. and Norman, G. (1988) *Building Britain 2001*. Centre for Strategic Studies in Construction.

Bennett, J. and Jayes, S.J. (1995) *Trusting the Team: The Best Practice Guide to Partnering in Construction*. Thomas Telford.

Bennett, J. and Jayes, S.L. (1998) *The Seven Pillars of Partnering: A Guide to Second Generation Partnering*. Thomas Telford.

Bennett, J., and Peace, S. (2006) *Partnering in the Construction Industry; Code of Practice for Strategic Collaborative Working*. Elsevier.

Acknowledgements

The authors would like to express sincere gratitude to people and organizations that helped in the creation of our book. In particular we wish to thank Andrew Bellerby Managing Director of Tekla UK, for providing a series of BIM images from past projects and their kind contribution towards the colour printing of the book, Afra Bindewald and Sascha Schneider for providing the information for the case study in Chapter 9, and Brian Moone, Mace Business School Director, for providing the information for the case study in Chapter 10.

Preface

Construction management involves unique challenges. It has features which are similar to the systematic improvement of products and production processes which characterise manufacturing but it also has features more usually associated with the controlled innovation and creativity which characterise project based industries, such as software development. The distinctive characteristics of construction result from buildings and our physical infrastructure involving many different technologies. Some are based on very local industries, some depend on companies which operate nationally but an increasing number of construction technologies depend on global networks of organizations often with widely different approaches to business. All this is further complicated by construction projects having individual locations which inevitably throw up at least a few surprises.

This combination of challenges is not comprehended by general management theories. Yet historically these provided the basis of most construction management courses. The inevitable result is young construction managers quickly discover the ideas they have been taught do not fit the practical situations they face. They find it difficult to make sense of bewildering mixes of terms, responsibilities and roles. Eventually most learn from experience in one sector of construction how to work reasonably effectively but that provides a poor basis for working in other sectors.

Similar limitations characterise construction management research. Too much of what is published distorts the realities of construction to make it fit theories developed in other industries. Inevitably the results appear remote from practice which has created a gulf between researchers and practitioners. The subject needs a new foundation which is firmly grounded in the characteristics of construction. This book attempts to provide that foundation by proposing a theory of construction management which identifies the actions which help construction projects and companies to be efficient.

The theory and the practical guidance which flows from it draw on knowledge and experience from two generations of construction management. The authors between them have been involved with the leading edge of construction management from the earliest days of its emergence as a distinct profession and academic subject through to contemporary best practice. When Milan Radosavljevic and John Bennett met in 2008 they rapidly found common ground in understanding the need for construction management to have a robust theoretical basis. They both recognised the absence of this essential foundation results in too much practice and far too much research being based on individual ideas and isolated initiatives. As a result good ideas are lost; systems to ensure year-on-year improvements in performance are weak or non-existent and progress in practice and research is painfully slow.

The fundamental aim of this book is to provide a basis for construction management to develop systematically on robust theoretical foundations. Theory is essential for practice and research to make the steady, relentless progress which is the hallmark of all outstanding industries and bodies of knowledge.

Given this high ambition, the book is organized to provide a coherent message for construction managers at all levels. It recognises that students and practitioners have different needs by developing the material in four sections, each designed to match the knowledge and experience of a distinct group of readers. In addition the authors have recognised the needs of the ever growing number of international students who come from different cultures and are not familiar with English construction terminology. The book therefore carefully defines all the key terms needed to understand the theory of construction management.

The book begins with a basic introduction to construction processes and products. This is in Chapters 1 and 2 and is suitable for first year undergraduate students in courses for all the professions involved in modern construction. The next section of the book describes the theory of construction management. It begins in Chapter 3 which defines the basic concepts of the subject. This is necessary because the construction management literature lacks consistent definitions of commonly used terms like *built environment*, *construction*, *design*, and so forth. Throughout the existing literature different terms are used for the same or similar concepts and the same terms are used for obviously different concepts. For example, a plethora of muddled and overlapping role titles are currently used in construction which makes it difficult to establish how projects are actually managed and by whom. Chapter 3 provides a set of clear and consistent definitions of the basic concepts needed to understand construction management.

The resulting set of fundamental definitions is used in Chapter 4 to describe the theory of construction management. This provides a rigorous way of understanding the factors which determine the performance of construction projects and companies. In a distinct break with most existing construction management literature, project and company management are treated as an integrated whole. This is vital in enabling the theory to take account of the major influence company managers have on projects, and the impact of project managers on companies. Chapter 4 also describes how the complexity and uncertainty endemic in construction can be expressed in mathematical terms to provide effective indicators of the inherent difficulty of the tasks facing construction managers in practice. The website linked to this book, www.wiley.com/go/constructionmanagementstrategies, includes a basic guide for readers not familiar with mathematical terms. The mathematics introduced in Chapter 4 is straightforward but nevertheless provides a powerful tool to guide decisions about appropriate strategies for construction projects and companies. The theoretical material in Chapters 3 and 4 is designed for undergraduate students in their final years as they become familiar with construction.

Chapters 5 to 9 describe the practical implications of the theory in the major construction management approaches currently used in practice. These include traditional approaches, the various management-based approaches as well as recent developments designed to foster cooperation including partnering and strategic cooperation. The book goes further in describing a totally integrated approach capable of delivering, in the right circumstances, outstanding performance. Each major approach has its own chapter which describes the main roles and actions and relates them to the theory of construction management. This rich mixture of theory and practice is designed for final year undergraduate students.

The first nine chapters are ideal for postgraduate students who have not studied the subject at undergraduate level. They provide a coherent and rigorous description of construction management in theoretical and practical terms. The subject matter is expressed in clear descriptions, diagrams and mathematics to make it accessible to the widest possible range of postgraduate students.

The first nine chapters provide an essential introduction to the fourth section which comprises Chapters 10 and 11. Chapter 10 describes how the theory of construction management benefits practice by providing 25 propositions about construction management actions which improve the efficiency of projects and companies. These are set out and explained in Chapter 4. The propositions provide a checklist of best practice. Practitioners who decide to act on any of the propositions will find a mass of useful advice and guidance on the strength and application of each of the propositions in the body of the book. As they consider using any of the major approaches to construction management, they will find the chapter which describes it helps ensure they are making a good choice and provides direct advice on using it effectively. All this is brought together in Chapter 10.

Chapter 11 describes the implications of the theory of construction management for future research. It then uses this analysis to propose a radical new basis for construction management research. It explains how this can be set up and developed by the construction management research community in a manner which enables individual projects and companies to use the best available knowledge and research. At present too much practice and research is isolated so that knowledge remains fragmented and lacks a robust basis for making progress with any confidence. Chapter 11 is intended to change this by proposing a major step forward for the subject. This important development is supported by the website linked to this book, www.wiley.com/go/constructionmanagementstrategies, which demonstrates the use of the proposed new knowledge base for construction management.

The book is based on the authors' very diverse knowledge and experience. It also takes account of the best of the construction management literature by including in each chapter a list of *Further Reading*. This lists the most significant books and papers which are relevant to the chapter. This approach has been adopted to avoid interrupting the text with detailed references to the sources of particular ideas. The authors fully understand why references are essential in research reports but the book is a textbook for students and a guide and checklist of best practice for construction managers. The needs of these readers are best served by guiding them towards the most outstanding construction management literature not by interrupting their focus on understanding the subject with a multitude of references.

Milan Radosavljevic
John Bennett
The University of Reading, UK

Chapter One
Introduction and Background

Construction provides many of humanity's greatest achievements: Salisbury Cathedral (Figure 1.1); the Taj Mahal (Figure 1.2); Sydney Opera House (Figure 1.3); high rise buildings in Dubai (Figure 1.4); and incredible buildings in modern China (Figure 1.5). Construction gives us places to live, eat, sleep, work, play, entertain, worship and be cared for. It provides the basis for transport systems and sophisticated services which make modern living comfortable and efficient.

Buildings and infrastructure involve virtually every human technology which makes them the most complex of products. They include technologies like brickwork and carpentry, which have their origins in ancient times, technologies based on heavy machinery, many of which developed during the first industrial revolution, right through to highly advanced, modern technologies including the most sophisticated communication systems and intelligent materials. Ensuring this diversity of technologies is used effectively and efficiently requires highly skilled management.

This book provides a rigorous guide to the situations and decisions which face construction managers. It is based on extensive research into the most effective ways of managing construction. Much of this research has been undertaken by the authors but the book also draws on published research into all aspects of construction management. The most important sources are listed at the end of each chapter as further reading.

Practice and research have identified fundamental concepts and relationships which guide effective and efficient construction management. These are described in this book in the form of a theory of construction management because this allows the ideas to be applied to every kind of construction project. More than this a rigorous theory allows the ideas to be developed by practitioners as new situations arise and robust ways of managing them are developed. It also allows the ideas to be tested by academic research and confirmed or replaced by better management ideas.

A fundamental theory of construction management needs to be based on a generic description which answers the question. What is construction? A useful way of providing such a description is to envisage visitors from another galaxy looking at Earth. This allows the description to be based on direct observation which is not influenced by preconceptions about construction.

Construction Management Strategies: A Theory of Construction Management, First Edition. Milan Radosavljevic and John Bennett.
© 2012 John Wiley & Sons, Ltd. Published 2012 by John Wiley & Sons, Ltd.

Figure 1.1 Computer Model of Salisbury Cathedral.

1.1 Construction viewed from space

As they circle Earth in their spacecraft the visitors from another galaxy see a planet covered by great expanses of blue water interrupted by land masses dominated by rocks and vegetation. The gleaming white polar ice caps attract their attention for a while. Looking closer at the land areas, the visitors see concentrations of buildings and infrastructure. In places these stretch for hundreds of miles forming mega-cities but most construction is arranged in smaller clusters which form cities, towns and villages. At night, the visitors see the Earth dominated by the lights of urban areas. They are fascinated by the erratic patterns of fixed and moving lights. They notice strings of lights connecting many of the cities, towns and villages.

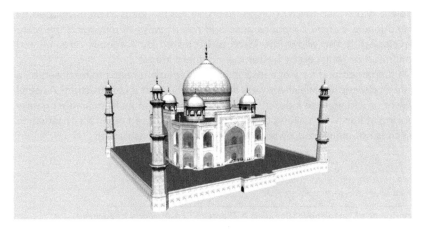

Figure 1.2 Computer Model of the Taj Mahal.

Figure 1.3 Computer Model of Sydney Opera House.

Figure 1.4 Computer Model of a Residential Area in Dubai.

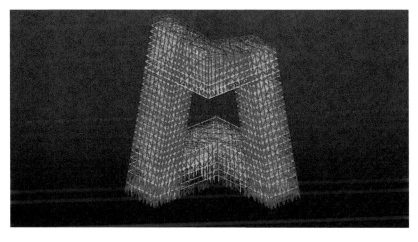

Figure 1.5 Computer Model of the CCTV Tower in Beijing. *Source*: Tekla Oy, Building Information Modelling software vendor

Looking at the same areas in daylight they see roads and railways carrying vehicles and trains. The patterns of movement cause them to notice concentrations of aeroplanes taking off and landing near cities. They see huge ships leaving and arriving at many of the cities near the oceans. Looking ever closer they see people in the urban areas moving in and out of buildings, walking between them and using various forms of transport.

As they focus on the urban areas, their attention is attracted by sites where new structures are apparently growing out of the ground. This growth takes time and involves people and machines in many different actions.

Observing a construction site our visitors see a group of people communicating and performing a complex set of actions that collectively contribute to a growth of a new structure. They are fascinated by huge excavating machines ripping earth and subsoil apart and pushing it into new, unnatural shapes. On other sites they see great tower cranes lifting materials and components into place. Some are forming massive steel frames. Others are lifting prefabricated concrete units to form the structure of a building, bridge or some other brainwave. Yet others are lifting prefabricated cladding panels and internal elements of buildings. Other sites are dominated by reinforced concrete technology as wooden or steel formwork is filled with reinforcement and concrete which is pumped into place from vehicles largely comprising huge, revolving tanks. Looking closer they see that not all construction technologies depend on big machines. There are groups of people who undertake actions which rely on their own physical strength and skills to position and then fix materials and components. In total the visitors see people using a wide variety of tools, equipment and materials.

Our visitors notice people on construction sites work according to day and night intervals; and in many cases they also see a pattern of work stopping for two days at regular seven day intervals. Initially they assume most groups of humans are working at different tasks but they may well see groups undertaking the same activity in different parts of the structure. They may guess these have some extra relationship beyond their involvement on the same site.

As they watch different examples of these fascinating sets of actions (Figure 1.6), the visitors realise they move through stages dominated by distinct types of technology. Before any construction starts the site may be an empty space or it may contain existing structures. The first stage alters or demolishes any existing structures and reshapes the site. This prepares the site for the next stages which create a strong foundation and a basic structure. The visitors can see broad similarities in the function of the foundations and basic structure but as they look more closely it becomes apparent that they have individual characteristics. Many different sizes and shapes are formed from various combinations of concrete, bricks, steel, other metals, timber or various synthetic materials. Some basic structures sit just above ground level or even below it but others provide many floors rising high into the air.

Once the basic structure is complete, the next stages clad it with various materials and components. As this external cladding is completed, further stages begin in the newly created internal spaces. Pipes, ducts and wires are threaded through the basic structure. Sometimes these are installed in large prefabricated units but equally often they are positioned by what appear to be specialists working with hand tools. Further stages form partitions to divide each floor into separate spaces. At the same stage various kinds of access

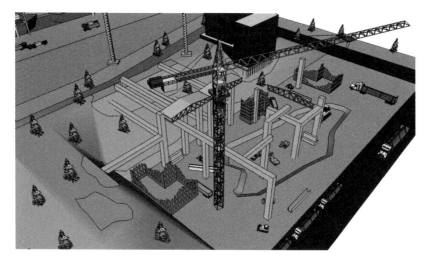

Figure 1.6 Construction in Progress.

between the separate floors are installed. These may be assembled on site from basic materials or involve the installation of complex components. Further stages install major items of plant and equipment which the visitors learn are designed to heat or cool the completed building, provide electricity in a controlled form, supply water, gas and other useful chemicals and dispose of waste material. The visitors see these various services form systems which are tested and re-tested to ensure they work properly. As all these actions are completed, further stages provide internal and external decoration to complete the new building or addition to the infrastructure.

The visitors recognise that on each separate site they are watching a concerted effort by a group of people to construct a new structure until a point in time when all the people involved in construction leave and are replaced by another group of people who use the newly created structure. In its most fundamental form our observers describe the construction actions as a complex interplay of people, tools, equipment and materials coordinated by communication.

As the visitors continue looking at many construction sites, they learn that humans refer to sets of linked actions which have agreed start and end dates as actions and projects.

As the visitors from another galaxy struggle to understand what they have observed they notice the actions on site are not independent. They see lorries and vans delivering materials, components, equipment and machines to the site. As they track the lorries and vans they recognise they are part of complex supply chains which link warehouses, factories, processing plants, mines and many different kinds of transport.

As they attempt to make sense of these wider patterns of actions (Figure 1.7), the visitors notice some of the lorries and vans are decorated with distinctive images which the humans call logos. They see the same types of logos on some of the warehouses, factories and the other parts of individual supply chains. Then following the materials and components onto construction sites they recognise groups of people performing distinct actions wearing clothes some of which carry the same logos. Other groups using different materials,

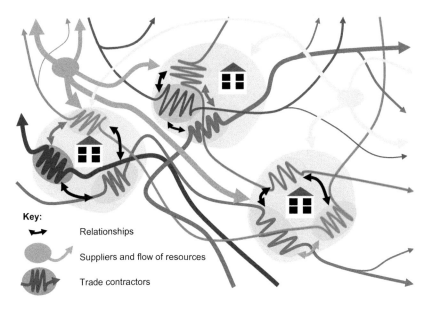

Key:

▶——▶   Relationships

⬤↗   Suppliers and flow of resources

⬤〰   Trade contractors

Figure 1.7   Construction Project at the Hub of a Complex Pattern of Supply Chains.

components, tools, equipment and machines are dressed differently and have different logos.

As they watch more sites, the visitors realise there are groups on other sites wearing the same logos. The visitors conclude that the logos serve to identify the existence of distinct organizational entities which humans call companies. As they consider this evidence, the visitors realise companies provide some kind of connection between people and the resources they use which is independent of individual construction projects. The visitors soon work out that individual companies provide the resources needed to carry out a specific type of production work on many different sites. By studying further, they find some companies operate locally or nationally, while others work all over the world.

Looking inside construction companies the visitors see they are permanent organizations intended to continue long-term. They listen to meetings of people called managers as they make decisions about the actions of the company as a whole. The meetings discuss staff training, investments in new plant and equipment, financial issues, developments in the demand for construction and new government legislation. All these issues concern the company as a long-term enterprise and are entirely independent of any individual projects.

Seeing the confidence with which managers in many of the companies deal with broad ranging issues makes the visitors from another galaxy question why the organizations responsible for construction projects are temporary. They have watched project organizations being formed and re-formed as projects progress through their different stages and then once the new building or infrastructure is complete, they cease to exist. The visitors find this puzzling but are unable to work out why people assemble project organizations only to disband them after just one project. It must be more sensible to let a carefully developed and efficient organization undertake more projects.

As they struggle to understand project organizations the visitors realise the sets of actions which make up most construction projects are so complex there must be a sophisticated system of coordination to ensure the work is undertaken correctly. Watching closely the visitors' attention is attracted by people called foremen who do not perform any production actions but communicate with those who do the work. Then the visitors notice other people, who humans call designers and managers, who communicate with the foremen. In addition to day-to-day, informal communication, there are formal meetings. These bring together various groups of people involved in the project at regular intervals to discuss problems and make decisions. The visitors notice the formal meetings are arranged and run by managers. It soon becomes clear the communication at formal meetings leads to certain work being done. They notice other less formal discussions between managers and foremen and as they get close enough to listen they discover that much of this informal communication is needed to prevent clashes between groups undertaking closely related construction actions. In its purest form the observers would regard these various kinds of communication as management of people, tools, equipment and materials.

As they watch people on site communicating, the visitors see them referring to various paper-based and electronic documents. They realise many of these documents are not produced on site but arrive from various external sources. They see communication on site becomes most intensive when new documents arrive. These discussions are led by managers who also appear to control the distribution of the documents. The visitors gradually realise the management of the construction actions on site is guided by information provided by the documents.

The visitors also notice most of the documents carry logos similar to those on the workers' clothes. By tracing the documents back to the originating companies they recognise each new building or addition to our infrastructure begins with tentative ideas. These usually originate in organizations which are not construction companies. The visitors see these organizations are primarily involved in some activity other than construction and have decided they need a new building or infrastructure. The visitors are fascinated as they watch how ideas for new construction emerge and change. They see men in smart suits sat around large tables arguing about minor features of a new building. They watch formal meetings of various government bodies debating the merits of a new airport. They see many discussions inside customer organizations as staff try to understand the implications of a new factory or office building. The meetings and informal discussions eventually lead to an agreed description of what the customer organization needs.

The customer organization approaches a construction organization either during their internal discussions or when a decision has been made that a new facility is needed. This triggers design work and a multitude of calculations. The visitors see descriptions of the end product being developed in ever greater detail. Various ideas are discussed and documented before there is agreement on one design. This is developed by people working in many different companies and results in detailed descriptions of all the parts of the new facility. The visitors recognise much of this detailed work is undertaken by companies which form part of the supply chains for construction projects. Other specialists consider how the emerging design can be constructed, how long it will take to complete and what it may cost.

As the visitors watch further, they see the formal meetings, informal communications and documents provide information which helps coordinate the design and management actions. They see managers guiding this coordination system. They realise the site and supply chain actions they have already studied include a similar system. Indeed they can see many projects managers use a common coordination system for all the actions whether they are based on site or elsewhere.

Looking back at their observations of the coordination systems, the visitors notice that managers spend much of their time on a day-to-day basis dealing with problems. Dramatic examples arise when a construction site is affected by bad weather. The visitors are amazed and amused by the chaos which follows snow, heavy rain, cold weather or high winds. In some parts of the planet, all the actions on construction sites are brought to a shuddering halt for many days by these extreme weather conditions. They watch managers struggling to find ways to protect the partially completed work and ensure an early resumption of effective work.

The visitors also remember being fascinated by construction sites that had became the subject of protests. They had watched people, many carrying banners and shouting, surrounding a site. The visitors witnessed protests which objected to the way the work was being organized and others where the protestors disliked the nature of the new facility. They saw protests provoked by sites working at night, streams of heavy lorries on narrow roads, the construction of a nuclear power station, a prison and a motorway which threatened to destroy the habitat of a spotted toad. They also noticed protests by construction workers about conditions on particular sites or their wages being reduced. Whatever the causes, it is plainly obvious that managers face many difficulties as they struggle to deal with protests and ensure efficient work on site.

As they continued discussing the problems faced by construction managers, the visitors identify a less dramatic but far more common cause. Individual construction actions often overrun their planned end dates. Many reasons and excuses are offered to explain these failures: shortage of materials, absent workers, broken machines, damaged components, work delayed by other people working on the site, an industrial dispute at a factory manufacturing components for the project, and many more. Whatever the causes, delays leave managers to find some way of making up lost time or explaining to the customer organization that their new facility will be completed late.

The visitors from another galaxy decide construction is complex and inherently uncertain. The uncertainty may have its causes inside the project organization or result from interference from external sources. They begin to understand that construction management is difficult and the fact that many constructed facilities are completed on time is a substantial achievement.

Turning their attention to the documents from several projects, the observers see some of the information which guides management contains values in a single or sometimes several currencies. They read that companies undertake work only if these sums of money change hands and humans generalise these transactions into economic principles. They also see the financing of many construction projects can be a complicated business as customer organizations attempt to borrow money from banks and speculators, and seek subsidies or grants from official agencies. They recognise the success of some construction projects depends largely on the terms and conditions accepted by the customer organization in order to obtain the necessary finance.

As they read further the observers discover the transactions are governed by documents called contracts. Looking through contracts, the observers see various clauses govern the relationships between the separate companies, the flows of information between them, and the work that needs to be completed in order to complete a new facility. They notice that on each individual project the customer organization is a party to a number of the contracts. Our observers read that at the end of projects, the customer organization takes over the new facility. In many cases they see the customer use the new structure to support their own actions but this is not always the case. Some new facilities are used by other organizations and these arrangements give rise to yet more contracts.

Further investigations of contracts and all the associated documents reveal the existence of another kind of documents which influence the actions of those involved in construction projects. These very formal documents are produced by organizations external to any of the companies involved. The visitors discover the external documents are called laws and regulations. They discover they are produced and published by various levels of government and other organizations working for government. In this way they identify that all actions, including construction, are governed by a legal system. One effect which intrigued the visitors is the preliminary stages of many projects are delayed by a need to obtain official approval for the particular type of facility required by the customer organization, the proposed design, particular design details or the planned method of working.

The visitors from another galaxy conclude that construction on Earth takes place in complex environments (Figure 1.8) which may interfere with even the most carefully devised strategies and plans of experienced construction managers.

Returning to their own planet the visitors' report is greeted with astonishment and laughter. Construction on Earth is very different to their own construction methods which allow individuals to make plans, consult with everyone likely to be affected, reach agreement on what should be produced and then place a firm order. The new facility is produced by robots using intelligent materials and never takes more than four weeks to complete. It will be several centuries before construction on Earth achieves this highly developed approach. In the meantime this book provides a guide to current best practice and the immediate future.

1.2 What is construction?

The visitors from another galaxy provide an independent view of construction based on direct observation. This provides the basis for a robust answer to the question: What is construction?

Construction is a series of actions undertaken by construction companies which produce or alter buildings and infrastructure. Individual construction companies become competent at one or more of the actions over many years. They apply their specialised skills and knowledge on construction projects. Each construction project has a start and end date and usually requires a number of construction companies that work together to produce a new or altered building, a group of buildings, or an addition or alteration to the infrastructure.

The actions which form any one construction project are extremely diverse as they take place in widely different locations and may involve practically every technology yet devised by humans. They include design and management

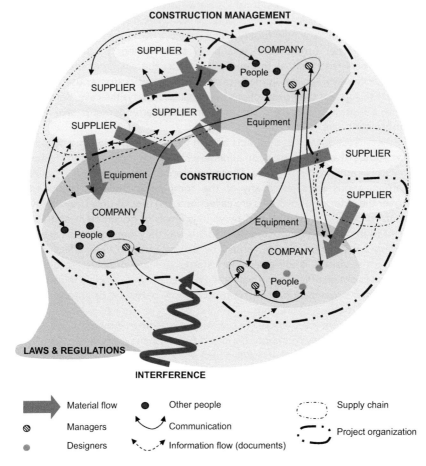

CONSTRUCTION MANAGEMENT

Material flow	● Other people	Supply chain
⊘ Managers	Communication	Project organization
● Designers	Information flow (documents)	

Figure 1.8   The Complexity Faced by Construction Management.

decisions which involve owners and customers in detailed negotiations with construction specialists and a great variety of regulatory, legal and financial organizations. These actions precede the direct physical production of the new facility which takes place on a construction site and in many distinct supply chains.

Ensuring the actions are undertaken effectively, efficiently and on time is construction management. It requires the coordination of a complex interplay of people, materials, components, tools, equipment and machines subject to variable performance in environments likely to interfere with planned progress. This in turn requires effective communication and efficient systems to organize the flow of the documents which provide the information needed by everyone involved in construction.

Construction management is the responsibility of everyone involved in construction companies and projects. It is common within companies and projects for specialists in management to be given responsibility for parts of the overall construction management task. This does not remove responsibility from everyone involved for ensuring that all the actions are undertaken effectively, efficiently and on time.

1.3 Why a theory of construction management is needed

Construction management is needed to ensure the specialist actions needed to produce modern buildings and all the parts of our incredibly complex physical infrastructure can be undertaken efficiently. As described earlier in this chapter, construction projects require a bewildering range of resources, knowledge and skills. They require finance, creativity, science, technology, architecture, engineering, factories, craftsmen, labourers and distribution systems aided by almost every type of machine. The number of construction companies directly involved in major projects runs into hundreds; while the supply chains which provide all the necessary materials, components and systems involve literally thousands.

Modern buildings and infrastructure are the most complex things produced by humans. Professions, trades and crafts have developed to play distinct parts in the challenging work of producing them. Some specialised construction knowledge and skills, such as that used by architects, bricklayers and carpenters, have existed for centuries and are widely recognised and respected. Others are very new. Indeed, it is the case that every decade sees the emergence of new disciplines to meet challenging new demands from society and individual customers.

Construction management emerged in the second half of the twentieth century as a distinct discipline. It exists to ensure that construction projects are completed efficiently. This is a demanding task because the sheer complexity of modern construction and the potential for project environments to interfere with even the most careful plans make construction inherently difficult. Construction management seeks to minimise this inherent difficulty in ways which allow projects to be completed efficiently.

Initially the new profession concentrated on providing effective management for individual construction projects. This reflected the project focus of the established construction professions. Construction's historical project focus results from the practical characteristics of construction. It produces buildings and parts of the overall infrastructure which are fixed in one location in response to the needs of individual customers. In total, additions and alterations to the built environment may be required almost anywhere on Earth. This means construction is faced with an endless stream of new situations and new challenges. Each construction site is unique, and each customer has distinct needs and demands. These fundamental characteristics have tended to dominate the thinking of those involved in the industry and this gives traditional construction practice a strong project bias.

The focus on individual projects has influenced the development of distinct professions, working methods and much construction research. This has led to a steady improvement in construction's efficiency and the quality of buildings and other construction products. However, in recent years it has become apparent that the focus on project management has limited construction's performance.

World class progress and development requires the construction industry's understanding of project management to be married to an equal focus on company management. Construction companies need to think and plan long term if construction is to become a truly modern industry able to meet the demands of the twenty-first century. Construction management is at the centre of both company and project management. This new understanding is vital for the construction industry's prospects. In simple terms, it is beginning to be recognised that construction projects can be undertaken efficiently and predictably only by well run construction companies.

Recognising the equal needs of projects and companies represents a major step forward in construction management thinking. In the past too much emphasis has been given to projects in the education, training and work of all the construction professions. This bias largely came from the practical need to focus on satisfying the needs of individual construction customers. However, many of construction's major customers are leading the way in demanding a long-term approach to the difficult task of producing major buildings and infrastructure. The leading edge of construction practice is already responding to these new and more challenging demands.

The emerging knowledge driving this leading edge are described in this book in theoretical terms because the most robust way of capturing human knowledge is to express it as a rigorous theory. This provides a tool kit of concepts and relationships which enable practitioners to analyse individual situations and select appropriate actions which fit their individual needs. In this way the theory provides a robust basis for everyone involved in construction to make decisions which improve their efficiency and the quality of construction products. A central assumption of the theory is that projects need to be well managed and companies need to be equally well managed.

The concepts needed by the theory of construction management are described in Chapter 3. This provides the basis for the theory to be stated in Chapter 4. The theory is then used in Chapters 5 to 9 to describe all the well established ways of managing construction projects. These chapters provide the background for Chapter 10 which describes the practical implications of the theory for current best practice and suggests future developments. Finally Chapter 11 describes how the theory can be tested and developed by research and practice.

The book is a textbook for all levels of students involved in construction. The theory helps all entrants to the construction industry understand the situations and challenges they find as they begin their own careers. Beyond that, the concepts and relationships described in the book provide a robust basis for understanding and developing best practice in all the construction professions.

1.4 Who can manage construction?

The various construction professions have strengths and weakness and once again our visitors from space can take an objective view of the character of key roles in construction. As they learn more about construction, the visitors may well discuss who should play the leading role in managing projects. They are likely to imagine various possible scenarios for construction management, some of which will be similar to those found in practice but others which are unlikely. In creating their scenarios our visitors will need to recognise that constructing a new facility, at the most generic level, comprises three distinct phases.

1. Preparation for construction on site (In this preliminary phase the facility has to be envisaged by a customer who orders the new facility. They employ various construction professionals who design the facility and manage the whole project.)

2. Construction in factories and on site (This phase is dominated by the actions of construction companies who usually have a formal contractual relationship with the customer. As a result many are called contractors. They arrange the supply of materials and components, and build the facility. This complex process normally requires managers to ensure it is completed efficiently.)
3. After construction is complete (Once the construction companies have finished their work, the customers, occupiers and facilities managers take over and use their new facility.)

Thinking about each of these challenging tasks will lead the visitors from space into discussing various possible scenarios for construction management. The following possibilities serve to illustrate the character of one of the key roles in construction projects.

1. Construction may be managed by designers (possible justification: customers require designers to envisage the overall facility and all its detailed features; so designers could well have the most profound understanding of the new facility).
2. Construction may be managed by customers themselves (Possible justification: customers have an obvious incentive to ensure their construction is well managed because they have to finance it and their organization will have to live with the resulting new facility. This should ensure they display the greatest dedication to the task of delivering the facility efficiently.)
3. Construction may be managed by contractors (possible justification: contractors are likely to have a high level of practical knowledge about all the actions needed to construct the required new facility).
4. Construction may be managed by facilities managers (possible justification: the constructed facility will be operated for many years so it is of paramount importance that the facility is simple to operate and maintain).
5. Construction may be managed by independent project managers (possible justification: project managers have a good overview of the whole life cycle of the project and so may be best placed to ensure customers get the greatest overall value from their new facility and its subsequent use).
6. Construction may be managed by independent construction managers (possible justification: by managing all construction actions efficiently, construction managers may be best placed to ensure customers get the greatest overall value from their new facility).

Identifying the advantages and disadvantages of each of these scenarios will help understand each of these six key roles.

1.5 Construction managed by designers

Designers work closely with their customers in deciding exactly what kind of facility is needed. They design the new facility and therefore provide an obvious choice to manage the whole project. However, their knowledge tends to be focussed on the performance and use of the facility itself. Indeed the designer's task is distinctly different from the challenges posed by manufacturing and production. The specialist knowledge they call on is very different from that needed for the direct physical actions required to produce the new facility.

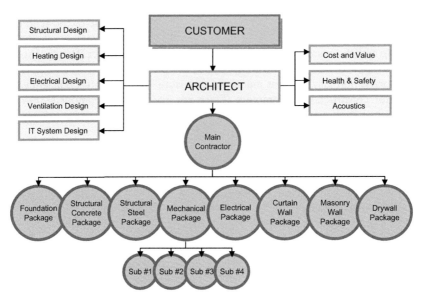

Figure 1.9 Designer-Led Construction Management.

Their lack of operational experience in organizing construction sites and all the supply chains which provide the materials and components could well lead designers into selecting inappropriate construction methods resulting in an inefficient use of resources. In this first scenario designers will need to go beyond their natural design task and extend their knowledge to the manufacturing and production processes. More importantly they need to establish effective relationships with all the construction companies that form the supply chains for materials and components and undertake the direct production of the new facility. They will need to do this and maintain effective relationships with the companies which provide specialist design advice. Figure 1.9 shows the general form of a designer led project organization.

Advantages	Disadvantages
Single point of contact for the customer Manufacturing and production guided by complete design information Designers being involved in all stages of the project may help ensure a high quality facility	Contractors involved too late to ensure the design takes account of the need for efficient manufacturing and production Designer may select contractors with an inadequate track record Problems caused by the design may create tensions between designers and contractors Designers lack of knowledge of manufacturing and production processes may cause inefficiencies The project may well be subject to regular design changes which interrupt efficient manufacturing and production

1.6 Construction managed by customers

Customers are normally the most committed party in the construction process. They will own and often use the new facility once it is complete. This means they have serious incentives to ensure the right type of facility is constructed at the right time, for the right budget. Their main weakness is they are in a different type of business from construction. While they may show high levels of dedication to managing their project, their inadequate knowledge of construction technology, resources, materials and components are major barriers to them being successful. This is compounded in most cases by a complete lack of established relationships with the construction supply chain. This can be a serious barrier because the construction of many facilities requires the close involvement of specialist teams in creatively discussing highly developed knowledge and working methods based on many years successful experience. Most customers do not have direct access to such expert, integrated teams. This is not surprising because many customers need only one new facility in several decades so there is no sensible reason for them to maintain teams that have sufficient knowledge to be able to directly manage the construction of a new facility. However, some customers do require several new facilities each year and may well decide it is justified to employ construction management capabilities in-house. They may recognise that having a deep understanding of the construction process is of crucial importance for improving their own products or services. In such cases customers may well have their own construction management teams. This situation is used to illustrate customer-led construction management and the general form of the resulting project organization is shown in Figure 1.10.

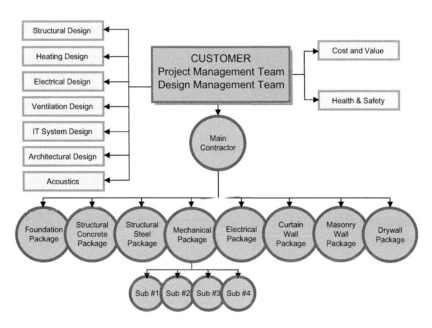

Figure 1.10 Customer-Led Construction Management.

Advantages	Disadvantages
Customer's dedication and involvement in the project should ensure the facility provides good value	Customer's business priorities may clash with the needs of construction and cause misunderstandings
Internal teams responsible for managing contracted organizations can ensure the customer's needs remain paramount	Lack of experience in working with contractors may create an inefficient project organization
Early involvement of contractors may ensure designs take account of manufacturing and production issues	A hands-on approach by the customer may result in clashes with contractors especially if they are employed on the basis of fixed-price contracts

1.7 Construction managed by contractors

Contractors undertake the direct physical production of the facility; they understand the direct physical construction process in detail and consistently work with specialist supply chains. This suggests they could provide excellent management for projects but contractors without design capabilities may face difficulties when trying to convert the facility envisaged by the customer into reality. Also contractors that undertake major projects are normally not involved in the operation and maintenance of the built facility which may limit their ability to produce a facility of the highest possible value to the client. For this scenario to work contractors either need a strong in-house design team or they have to subcontract design to an external organization they can work with effectively. Subcontracting provides a major obstacle to the success of this scenario. If a main contractor subcontracts the design and major work packages then their role is reduced to contract management which gives them very little power to ensure the efficiency of the construction process. Nevertheless, some global contracting organizations have build up strong in-house capabilities with large in-house design, specialist production and facilities management teams. Such companies subcontract only a few minor work packages and have competent in-house teams able to undertake all the major construction actions. This is one of the most appropriate construction management approaches. Figure 1.11 shows the main features of a contractor led project organization.

Advantages	Disadvantages
Single point of contact for the customer	Only large contractors have adequate competencies and capabilities to avoid extensive subcontracting
Potential for minimising the number of teams in the project organization	Where the main contractor subcontracts major packages and design, the project organization is fragmented which tends to cause difficulties

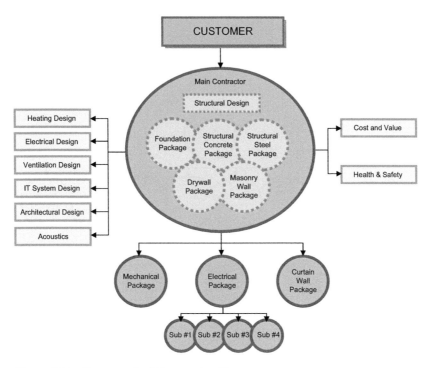

Figure 1.11 Contractor-Led Construction Management.

Advantages	Disadvantages
Experienced management of construction teams Established relationships within the supply chain	

1.8 Construction managed by facilities managers

Operation and maintenance account for a bigger proportion of the whole life costs of constructed facilities than does the initial construction. It is therefore not surprising that whole life costs are increasingly given considerable attention by customers and construction companies. In part this is driven by increasingly stringent legislation aimed at ensuring construction is sustainable. Surging energy prices are adding to the importance of constructed facilities being operated and maintained efficiently. One significant result is customers are involving facilities managers in the construction process as they are best placed to determine the most appropriate type of technology and construction methods to achieve minimum operational and maintenance costs. This expert focus on minimising the costs of operation and maintenance often produces high quality facilities. However, going further and requiring the facilities

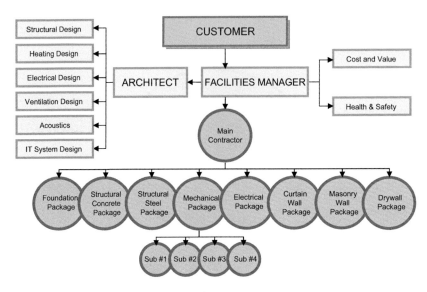

Figure 1.12 Facilities-Management-Led Construction Management.

manager to undertake the construction management role is likely to raise problems. Facilities managers lack direct experience of the whole construction process. They are likely to have little understanding of the design, manufacturing and production processes on major projects. Furthermore, the nature of their task does not allow them to develop effective relationships with specialist contractors. These weaknesses mean this scenario is unlikely to be a viable option for efficient construction management. It is more likely that facilities managers will continue increasing their involvement in construction projects but only during the inception, design and commissioning stages. Never-the-less, it is instructive to consider how facilities-led construction management might work (Figure 1.12).

Advantages	Disadvantages
The facility is likely to be efficient in terms of its whole-life costs	Lack of experience in working with designers and specialist contractors may well result in an inefficient project organization
Contractors involved in the construction process may subsequently be involved in the operation and maintenance of the facility	Contractors involved too late to ensure the design takes account of the need for efficient manufacturing and production

1.9 Construction managed by independent project managers

Independent project management organizations are often employed to manage complex projects through many stages in which construction is only a

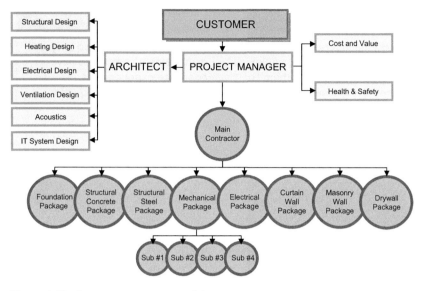

Figure 1.13 Project-Management-Led Construction Management.

relatively small part. Many of these organizations are generalists relying on huge supply chains which give them diverse knowledge. This is likely to give them a good overview of the issues and potential benefits which persuaded a customer to invest in a new constructed facility. However, if they were to undertake the management of a construction project, their wide ranging knowledge could all too easily leave the project with widely conflicting objectives. Similarly an over reliance on contractual relationships may limit the development of the construction project organization because of the project management organization's reliance on rudimentary contract management rather than much more effective relationship-based management. A generalist project management organization may also lack the specialist knowledge needed to give adequate consideration to appropriate construction methods and an effective focus on construction supply chain management. There are project management organizations which specialise in managing construction projects and understand the need for well informed management. The existence of these specialist organizations are taken into account in illustrating this scenario (Figure 1.13).

Advantages	Disadvantages
Management by project managers with experience of diverse temporary project organizations Project managers provide a range of views based on their close relationships with a number of large customer organizations	Project managers may employ subcontractors to provide at least some of the essential knowledge and skills leading to an over-reliance on contractual relationships Project managers may be oblivious to design issues because their normal role requires a wider and more generalist approach

Advantages	Disadvantages
Project managers experience of all the stages of large, complex projects may enable them to generate greater value for customers	Only large and established project management companies have adequate experience of manufacturing and production processes
Specialist construction project managers can provide effective and highly focussed management	Contractors involved too late to ensure the design takes account of the need for efficient manufacturing and production

1.10 Construction managed by independent construction managers

Many experienced customers rely on professional construction management organizations to manage their construction projects. In this scenario construction management organizations do not undertake any direct construction actions and all the designers and specialist contractors have contracts directly with the customer. The construction manager concentrates on managing the construction project organization on behalf of the customer. Solid relationships built up through many projects and possibly over many years are of vital importance for the success of this scenario because all the relationships between specialist teams are based on contracts. Construction management organizations of this type have very good knowledge of the construction process based on long experience and extensive experience in developing effective relationships. This scenario offers one of the best construction management approaches when it is based on solid long-term relationships. Even with this advantage, construction managers need to remain closely involved throughout their projects. It is all too easy for any problems not dealt with quickly to fester and lead to inefficient and dispute-ridden attitudes and behaviour. Construction managers need to concentrate relentlessly on ensuring everyone involved actively fosters effective relationships. Figure 1.14 shows the general form of a project organization led by an independent construction manager.

Advantages	Disadvantages
The construction manager's influence on design provides a robust basis for manufacturing and production	Design may be compromised by an over emphasis on manufacturing and production issues
Construction manager is likely to have long-term relationships with highly effective contractors	Over-reliance on contracts since every specialist contractor has a contract with the customer and this may lead to problems and disputes
Efficient manufacturing and production processes	

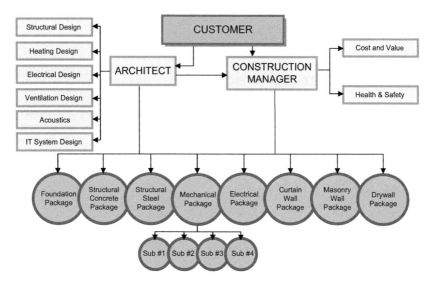

Figure 1.14 Construction-Manager-Led Construction Management.

1.11 How the construction industry works

The six scenarios which our visitors from another galaxy may have discussed provide opportunities to think about how some of the main roles in construction interact. It is now time to consider the approaches which do in fact dominate established practice.

All the leading national construction industries have developed over centuries. Distinct methods and procedures have grown up to serve the interests of customers, industry and the wider society more or less well. The main drivers for change are new technologies, new demands from customers or regulators and construction companies seeking more effective and profitable ways of working. Changes initiated by construction companies tend to be responses to economic crises or changes in market demand. Individual construction companies invest in research and development usually with financial support from government but it often takes decades for even the most useful industry-based innovations to be adopted widely.

As a result of these various factors, construction is organized in a number of distinct ways which produce significant differences in established national approaches to construction. These provide distinct strengths and in particular situations have various weaknesses. The leading edge in most national construction industries is dominated by large customers and construction companies. They lead negotiations with government about regulations and the way the public sector employs construction companies. Outside the leading edge are other more or less informal ways in which construction companies deal with their customers and work together on projects. However, the established national approaches are of most direct interest at this point in the book.

The next three sections describe the significant features of three influential approaches which provide an immediate introduction to the diversity of practice. Each is shaped by the interests of a different one of the main construction professions.

1.12 Designer-led practice

Traditional practice in the United Kingdom construction industry influenced practice in many countries as British methods and procedures were introduced during the days of the Empire. The traditional United Kingdom approach is dominated by designers. This is most evident in the building industry where architects have historically determined the methods and procedures used by customers and the other members of project organizations. Within this approach the customer for a new building employs an architect to help determine what is needed and then design it. As the design is developed, the architect recommends the customer employs engineers and other specialist designers who assist the architect in completing the design. The professional architects and engineers normally negotiate a fee with the customer taking account of the requirements of the project and their particular skills, experience and costs.

The resulting design team produce detailed drawings and specifications which provide the basis for selecting a contractor to take responsibility for the production work. The selection process normally involves competitive bidding and some negotiation over the precise terms of the contract. It is usual to appoint the contractor submitting the lowest price who enters into a contract with the customer to complete the work in accordance with the drawings and specifications by a specified date for an agreed sum of money. Quality is assured by requiring the contractor to be supervised by the architect. It is usual for the contractor to employ specialist subcontractors to undertake the production.

Most contractors have established working relationships with groups of subcontractors competent in the technologies normally involved in the projects they undertake. In selecting a subcontractor for each distinct technology, they may simply negotiate with one of their established subcontractors or invite prices from three or four and select the one submitting the lowest price. The architect's dominant role complicates these arrangements in various ways. Most significantly, it is likely that at least some members of the design team are companies whose main business is undertaking production in accordance with their own design information. These typically include companies specialising in structural frames, external cladding, internal partition systems and all the services required in modern buildings. Various arrangements exist to ensure the firms involved in the design are employed by the contractor or at least that they are considered for the work. This can lead to the contractor having to employ a company they have never worked with before; or even worse, employing a company they are in dispute with on another project.

The most significant of the distinct arrangements which have developed to facilitate the architect's detailed involvement in all stages of building projects is the creation of the profession of quantity surveying. The role of quantity surveyors is to prepare bills of quantities. These are detailed schedules of the work required to turn the architect's design into the finished building. Bills of quantities incorporate the design specifications and are sent with the main design drawings to contractors invited to bid for the work. Priced bills of quantities provided by the successful contractor form part of their contract with the customer. The unit rates in the priced bills of quantities are used to value changes to the design ordered by the architect as the work proceeds. This provides an administratively simple way of valuing changes which is reasonably equitable to customer and contractor provided the changes are relatively minor. Large changes lead to disputes, negotiations and occasionally result in litigation.

Quantity surveyors now use their knowledge of detailed building prices to provide cost advice to customers and designers during the design stages of projects. This helps ensure designs are likely to be capable of being constructed within the customer's budget. This constraint on design freedom is resented by architects and engineers who claim financial constraints prevent them producing great architecture.

Partly as a result of the complex and bureaucratic procedures which characterise traditional practice in the United Kingdom building industry, production work on site is unpredictable. Changes may be imposed at any time by architects who insist that every detail warrants careful design and if better answers can be identified, re-design. As a result work on site is planned on a day-to-day basis. It is characterised by delays and muddle which create inefficiency, uncertainty and disputes.

The focus on individual design limits the development of companies. Project organizations have to be assembled to undertake the particular set of actions required to realise each new design. Everyone involved faces new situations. Flexibility is crucial to survival and there is no advantage in companies growing big. Small companies struggle to find sufficient work which they are more or less competent to undertake. As a result the United Kingdom design-led approach gives little attention to company management which means long-term issues are neglected. The most serious effects are inadequate investment in training, research, development and marketing. This failure inevitably produces a weak industry. Many projects are completed late and over budget. Quality is generally acceptable but minor items of work are often incomplete when buildings are handed over to customers. The approach provides some of the world's finest architecture but it is just as likely to result in mediocre designs. United Kingdom urban areas are a strange mixture of high quality and great architecture surrounded by much that is mediocre and dull.

Overall the United Kingdom's design-led approach produces variable results in a largely unpredictable manner which customers increasingly find unacceptable. This has led in recent decades to many initiatives aimed at improving the industry's performance. The best seek to retain design quality but manage it in ways which deliver greater efficiency and certainty for customers. Many of the most effective ideas have their origins in the influential approaches described in the next two sections.

1.13 Manager-led practice

The most distinctive approach to construction in the United States is used by developers who undertake major building projects which they see as financial investments first and architecture second. Their approach relies on market forces allied to efficient management. At the start of a new project, major developers employ a design team and a construction management team. Some developers select these key teams on the basis of competitive bids. Others, particularly if they need their new facility completed as quickly as possible, negotiate tough terms with designers and managers they have successfully employed on earlier projects.

The two teams work together to establish the customer's requirements and produce a design and construction plan. This can be achieved very quickly, once the necessary formal approvals have been obtained. Design teams comprise architects and engineers who aim to produce the best possible design within the money and time allowed by the developer. Architects understand how useable floor areas, rental values, building costs and financial markets interact to constrain design. They concentrate on the 'feel and face' of their buildings as viewed by users and the general public. The structure and services are designed by engineering consultants who understand the requirements of local authorities and developers. They rely on their local knowledge in defining the required performance of the main elements and systems but leave detailed design decisions to the specialist contractors.

Construction managers aim to ensure the building is completed on time, within the developer's budget. During the design stages they establish an overall strategy for the actions on site which aims to ensure the design can be realised by an efficient production process. The details of individual specialist contractor's supply chains and production methods are left for them to determine. The construction management team concentrates on establishing a management framework which coordinates the work of designers and specialist contractors.

The specialist contractors selected to form project organizations are those offering the lowest price for a work package which produces an element or system of the end product. Once appointed, the specialist contractors undertake detail design, manufacturing or organizing the supply of materials and components, and producing a distinct element or system of the building. They are expected to ensure their decisions fit in with the work of all the other specialist contractors.

This approach provides a very fast way of working which depends on each local construction industry having well- established technologies understood by designers and specialist contractors. This ensures, for each work package, there are many specialist contractors with well-developed design details and production methods which meet the requirements of local designers. It also ensures there are standardised components readily available from local builder's merchants which meet the needs of locally established technologies. These are supported by excellent product information and technical advice which is readily available over the internet. This well-developed and standardised approach allows competitive bids for work packages to be based on the design teams' generic descriptions of the required product. It also guarantees fierce competition for the work packages on major projects.

The design team coordinates the design work of the specialist contractors to ensure the overall design concept is realised. The construction management team coordinates the manufacturing and production actions of the specialist contractors to ensure the customer's budget and completion date are achieved. A vital weapon in this is a programme of key decision points agreed by each company when they are selected to undertake work on the project. The key decision points typically link the completion of a defined set of information or a distinct stage of the production work to a specific date. The start of each new stage is marked by what is often called a 'kick-off' meeting. It brings together everyone involved in the stage to ensure they understand the scope and objectives of the stage and their individual responsibilities. Projects are driven towards these key decision points by coordination meetings attended by managers from all the companies directly involved in the current stage of the project. The meetings are organized and run by construction managers and concentrate on checking progress and solving major problems. A positive, 'can-do' attitude builds team spirit and energises everyone to strive to meet the agreed objectives.

It is inevitable that problems arise given the market driven, fast-track approach insisted on by major American developers. The relentless focus on time and cost combined with having at least forty or fifty major specialist contractors involved makes it inevitable that clashes and incompatibilities between individual design details will emerge as the production work proceeds. Workers directly involved in the problem on site are expected to cooperate in finding a mutually acceptable answer. They are expected to do this immediately the problem is identified so work is not delayed and no additional costs are incurred. It is only problems which cannot be solved by the workers directly involved which are dealt with at project coordination meetings. This dual level management system enables most projects to be driven to a reasonably satisfactory completion.

The relative certainty provided by locally standardised design and construction methods mean companies know broadly what to expect from any new project. This means they can afford to develop systems which improve their efficiency and invest in marketing to ensure reasonably stable workloads. As a result some companies grow big enough to shape developers' demands, influence local standards and have a beneficial influence on the industry's performance. The most successful operate across the United States and internationally. The management of such companies balance their own long-term interests with those of individual projects.

The overall results are that the American management -based approach produces buildings quickly, usually on time and within budget. They look stylish but rarely provide great architecture. Superficial the quality looks fine but many details, especially those hidden behind the glossy claddings, are completed in the quickest and cheapest manner with no regard for their appearance. The buildings lack any real depth of quality but broadly they satisfy the demands of major developers. The visual sterility of many urban areas in America is the social cost of efficient standardisation. The approach is like a highly developed machine which performs specific tasks efficiently.

However, machines struggle when faced with work which is different from that they were designed to undertake. Similarly, the American, management-based approach depends absolutely on sticking to established designs. Faced with innovative designs, the approach struggles. Irregular shapes, unusual details or new materials take specialist contractors beyond their well-established competence. Problems arise and the relentless focus on speed and economy obstructs attempts to find well thought out solutions. Dangerous compromises have resulted in building collapses, time overruns, higher costs, everyone blaming everyone else for the failure and litigation. Allowing these things to happen is a failure of construction management.

1.14 Contractor-led practice

Traditional practice in Japan is dominated by major contractors who have adapted the methods employed by Japan's great manufacturing companies.

This involves taking responsibility for every aspect of the construction projects they undertake. As a result leading construction companies deliver excellent buildings and infrastructure totally reliably. The companies are dominated by engineers recruited from university and trained over many years in the company's way of working. Some engineers specialise in architecture and undertake the design of buildings. The companies have world-class research and development departments which concentrate on producing new design ideas and new technologies which enable them to steadily improve the buildings and infrastructure they produce for their customers.

Japanese major contractors have established long-term relationships with major customers based on delivering exactly what these customers demand. This is all achieved exactly on time and at the agreed price. Customers know what they can reasonably demand because reliable and detailed information about the performance, timescales and costs of new buildings and infrastructure is published widely in Japan.

The major contractors have developed wide-ranging competence in all aspects of design and construction management. As a consequence, when they receive a new order from one of their established customers, they immediately set up a project committee. Its members are drawn from all parts of the company likely to suggest improvements to the customer's requirements, the design or construction methods. They often involve the company's research and development department in searching for the best answers.

Once the broad nature of the project is agreed, a project manager is appointed. Typically this is an engineer who has worked in the company for at least fifteen years and fully understands how the company works. He ensures that every aspect of the project is considered, re-considered and agreed before work begins on site.

The major contractors have long-term relationships with specialist subcontractors who undertake the production work. Typically these specialist contractors have worked for the major contractor for decades, even centuries. They know the type of designs the contractor produces and are competent in undertaking the required production actions safely, to reliable quality standards, exactly on time for the normal price.

When work on site begins, the design is complete and the construction plan is very detailed. This is facilitated by the use of extremely detailed national standards which specify every aspect of established construction technologies. For example, national standards include comprehensive sets of structural steelwork connection details. This allows project drawings to be relatively simple and specifications to be brief since they deal only with unusual design details. The wide use of standards means subcontractors do not face new or unusual tasks. The project manager can be sure they will be able to work effectively and so it makes sense to plan for all the production actions before work begins on site. The plan is extremely detailed. It defines each day's work in detail. For example, it establishes a precise time and date for the delivery of all the required materials and components.

Work on site is controlled by a small project management team led by the project manager appointed at the start of the project. Work begins at 8.00 a.m. five days a week with a meeting of everyone working on site that day. The subcontractor teams line up in order and are addressed by the project manager who describes the day's work. He highlights the required outcomes, any major deliveries and any safety issues which need attention. By 8.15 a.m. all the workers are in their work places where each construction team holds a brief meeting to agree exactly how they will complete the planned day's work. Well before 8.30 a.m. purposeful work begins.

At 3.00 p.m. all the foremen currently working on site meet with the project team to review the day's progress. Any problems are described and solved. The meeting is direct and tough. The foremen take the initiative in insisting that any subcontractor failing to complete the day's work must take remedial action. This may mean a team working late, bringing extra workers onto site, or in extreme cases working through the weekend to ensure the week's work is always completed in the week.

As a result of the detailed planning and control, projects are always completed exactly on time. They are fully complete in every detail when they are handed over to the customer. Everything has been checked and tested to ensure every part of the building or new infrastructure works properly. On completion, there is no negotiation over the price; the customer pays what was agreed when they initially placed the order for their new facility.

Everyone involved in these impressive organizations is required to search for better ways of doing their work. This is an essential part of total quality control which is procedures and actions designed to ensure everyone relentlessly searches for better ways of working. Every individual team is required to report twice a year on the improvements to their own performance they have achieved in the preceding six months. Companywide competitions are held annually to identify and reward the best new ideas.

The Japanese government actively fosters and supports the construction industry. They work with the major contractors in determining public policy towards construction. The biggest contractors undertake all the largest and most important public sector projects. Public sector investment is used to offset fluctuations in private sector demand so the major contractors have stable workloads. The most important effect is that construction companies have the confidence to invest in their own future. This valuable outcome is reinforced by the contractors themselves employing large numbers of experienced staff in marketing aimed at winning new orders from private sector customers.

The overall results are high levels of efficiency and remarkably reliable performance. The resulting buildings and infrastructure are produced to high quality standards. Designs tend to be safe, even dull but they broadly suit Japanese culture which requires individuals to fit in. In Japan it is often said "the proud nail is hammered flat." Individually designed buildings are demanded by a few customers and the major contractors can deliver these but they put great strain on their highly tuned systems. The resulting buildings are completed on time to the normal quality standards. This requires extraordinary efforts by everyone involved and often leaves the contractor with a financial loss but no loss of face.

These impressive results are possible because government and industry work together to create the conditions needed for very large construction companies to invest long term in training, research, development and marketing. They recruit graduates from the best universities and provide comprehensive and well-organized training which ensures they are competent in the company's working methods. To ensure their long-term survival, they have large marketing teams of highly experienced, senior engineers working with their long-term and potential customers. Their task is to identify situations where new construction could benefit a customer's business. The research institutes, which are large and well equipped, play a key role in the large companies' success. They continuously

improve tried and tested technologies to ensure the buildings and infrastructure they produce are safe, of reliable quality and function well. They
also produce bold concept designs for such things as mile-high buildings
and cities under the sea. These are often featured on national television
which helps sustain the construction industry's good reputation in Japan.

1.15 Conclusions

This chapter introduces the nature of construction in generic terms; discusses the
character of key roles in the construction process; and describes three influential
approaches to construction projects which have distinct strengths and weaknesses. This completes the basic introduction to construction's methods and
procedures and the role of construction management. The next chapter completes the essential introduction by describing the built environment to establish
a common understanding of the purpose of all construction actions.

Exercise

Figure 1.15 shows a simplified example of contractor-led construction
management with most of the design, auxiliary services and some of the
major mechanical and electrical work packages being subcontracted to
specialist designers, consultants and trade contractors.
   Some major contractors may comprise several internal specialist teams.
Draw different simplified configurations of contractor-led construction

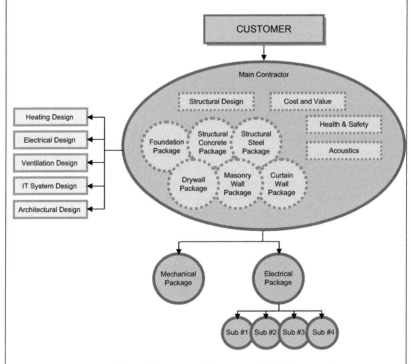

Figure 1.15 A Possible Configuration of Contractor-Led Construction
Management.

management. How many different configurations can you draw? Hint: Figure 1.15 shows one such possible configuration.

Take a close look at Figures 1.9–1.14 and draw possible simplified configurations for all the other construction management scenarios. How many can you draw? What does that tell you about construction management?

Further Reading

The following publications are the source of ideas used in this chapter and provide further information for readers.

Bennett, J. (2000) *Construction the Third Way: Managing Cooperation and Competition in Construction*. Butterworth-Heinemann. This book provides a wide-ranging description of key ideas which have shaped construction management practice and theory. It does so by explaining international best practice in managing construction in terms of fundamental ideas drawn from general management theory.

Bertelsen, S., and Sacks, R. (2007) Towards a New Understanding of the Construction Industry and the Nature of its Production. *Proceedings of the 15th Conference of the International Group for Lean Construction*, C. Pasquire and P. Tzortzopoulous (Eds.), Michigan State University, East Lansing, Michigan, 46–56. The paper challenges the current construction management approaches by looking into the nature of the construction industry as a complex network of projects that share the same production system and are therefore highly interdependent.

Cox, A. and Townsend, M. (1998) *Strategic Procurement in Construction*. Thomas Telford Ltd. This book recognises the consequences of the flawed presumption that there is a single approach that can be adopted regardless of the circumstances.

Morledge, R., Smith, A. and Kashiwagi, D. T. (2006) *Building Procurement*. Wiley-Blackwell, 1st edn. This book provides a good overview of the current procurement strategies starting with a concise and well-supported literature review. It defines core procurement principles and looks at its different aspects (e.g. procurement stages, risks, etc.). The book also provides a very good insight into construction procurement across the globe (e.g. Europe, China, and the United States).

Chapter Two
The Built Environment

2.1 Introduction

Humans have large brains probably due to a chance mutation some 200,000 years ago. Over the ensuing thousands of years humans have dreamt, imagined, innovated and made things. For most of this time, progress by today's standards was unbelievably slow but the last 10,000 years have seen an explosion in this creative enterprise. Looking at the results, it can reasonably be argued that the construction of the built environment is the most striking achievement.

This chapter describes the products of construction which form the built environment. It is not a single achievement but an accumulation of the efforts and enterprise of many generations. This is why the built environment provides such a wonderful variety of landmarks and sensations. These include justifiable famous buildings typified by the Coliseum (Figure 2.1) and the Pantheon (Figure 2.2) which have survived from ancient Rome. However, even more utilitarian examples of the built environment are impressive as a short journey from the Cotswold Hills to London, a little over 100 miles, beautifully illustrates. At first we see tiny villages of picturesque stone houses, often dominated by a church spire, looking as if they had grown out of the rolling landscape. The occasional group of farm buildings remind us of the significance of agricultural. Soon we pass Cirencester, the capital of the Cotswolds, which began as a Roman settlement some 2,000 years ago. In the middle ages, the town grew rich on the wool trade, declined in the eighteenth and nineteenth centuries and has now reinvented itself to provide a regional shopping centre and an elegant place to live. Heading south we pass through Swindon which grew up around Brunel's great railway works 200 years ago, declined as manufacturing moved overseas and was rescued by the building of the M4 motorway which enabled it to become a regional commercial and shopping centre. Joining the M4 motorway we head toward London. At first we pass through rural countryside and small towns but beginning with the university town of Reading, we see bigger sprawling urban conurbations. We pass Windsor, dominated by the Queen's magnificent castle Figure 2.3), and now enter the virtually continuously built up area which is greater London. We pass under the M25, London's outer circulatory motorway, and immediately become aware of Heathrow, claimed to be the busiest international airport in the world (Figure 2.4). The traffic is increasingly congested. The buildings are higher and closer together. We drive along raised motorways past thousands of

Construction Management Strategies: A Theory of Construction Management,
First Edition. Milan Radosavljevic and John Bennett.
© 2012 John Wiley & Sons, Ltd. Published 2012 by John Wiley & Sons, Ltd.

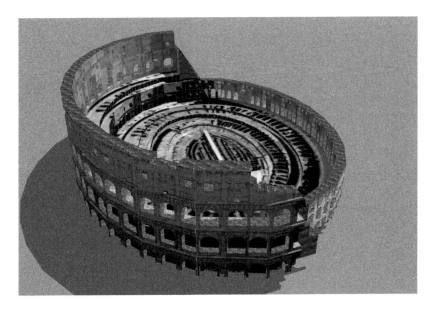

Figure 2.1 Computer Model of the Coliseum in Rome.

houses which form suburbia into the central districts of this great city. The M4 has ended and we are on the route of the old Bath Road, now a congested dual carriageway. We pass great museums and art galleries, smile at Harrods' amazing display windows, pass the green oasis which is Hyde Park before driving along the elegant Piccadilly and finally arrive in the totally commercialised Piccadilly Circus.

In a journey of less than 100 miles we have seen a few of the many different forms the built environment can take. If we travel further, cross frontiers and oceans, we see very different cities, towns, villages, buildings and diverse

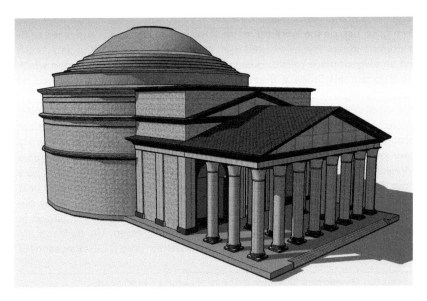

Figure 2.2 Computer Model of Pantheon in Rome.

Figure 2.3 Computer Model of Windsor Castle.

infrastructures. Some display the benefits of great wealth; many are dull, while others all too obviously result from deep poverty. There are districts which look vibrant and full of life and energy, while others have apparently been abandoned and left to rot. Most urban areas are formed from many different styles, sizes and shapes of buildings. They are often enlivened by parks, squares and other public spaces which mimic the countryside. They are chopped into districts by road systems and motorways carrying millions of cars and lorries. Railways carve equally indiscriminately over bridges and through tunnels into the heart of cities and towns. Airports and harbours link nations; border defences separate them.

This chapter describes the forces which shape the diverse built environment. Construction managers need to understand these forces because they

Figure 2.4 Computer Model of Heathrow Terminal 5.

determine what their construction projects are required to produce, set limits on the approach used in undertaking those projects and have the capacity to interfere with the most carefully prepared designs and plans.

2.2 Forces which shape the built environment

The most fundamental forces which shape the built environment are the basic human needs for shelter, safety and convenience. The practical effects of these forces have changed over time as ever more sophisticated ways of satisfying basic human needs have been devised. It is instructive to visit old buildings and consider just how primitive they are. Many people still suffer from having to live in poor quality buildings as simple as those preserved in museums. Indeed it remains an international disgrace that many slum dwellers endure conditions every bit as bad as those described in Charles Dickens' novels which shocked Victorian London. These however are rarely the direct concern of the construction industry. Slums are, by definition, neglected by the affluent majority.

Construction managers' direct concerns are modern buildings and infra-structure. They provide complete shelter from the normal climate in all parts of the globe where humans live in any significant numbers. Occasional freak weather conditions still cause disasters. Earthquakes, floods, lightening and high winds can damage buildings and infrastructure; and injure or kill people. These rare events dominate the news media for a few days but the vast majority of people in the developed world enjoy shelter which protects them, even from freak natural events.

A major concern in the modern built environment is the need to feel safe from other people. This may result in thick walls, strong structures, steel bars and shutters, robust door locks and formidable boundary walls and mighty gates. The direct physical barriers are increasingly supplemented by sophisticated technology including sensors which can detect illegal activities, lights which respond to movement, and cameras which record everything. People are safer than ever before in the developed world. One consequence is hysterical media headlines reporting occasional failures of modern safety measures which can make people feel less secure than is in any way justified by the facts.

The third basic requirement of the built environment is that it supports and helps people undertake the activities they wish to pursue. Historically buildings were not particularly convenient. Water had to be carried in buckets from a well and heated over a fire. Sanitary arrangements were primitive and often involved burying sewage in pits. The built environment has changed dramatically. Heating and cooling systems in large buildings feed air conditioning which can be tailored to the individual needs of the occupants of each separate room. Hot and cold, even boiling water is available on tap. Sanitary fittings enable us to wash, shave, shower or bath in comfort. Toilets not only dispose of waste matter but they can analyse it and diagnose the user's state of health. Light and electric power is readily available at the flick of a switch. Gas and other chemicals can be piped to wherever they are needed. Communication systems can be channelled to every room to allow users to talk and send messages all over the world in an instant. These levels of convenience are taken for granted in the modern built environment and construction managers need to check exactly what their customers assume will be provided in their new building or addition to the infrastructure.

In addition to these basic needs, the built environment is shaped by other forces. Some of the most significant are described in the next sections.

2.3 Climate and geology

Climate and geology directly influence the design and construction of buildings and infrastructure. The insulation of the external envelop of buildings, the use of sloping or flat roofs, the use of natural ventilation, access to buildings, road surfaces, the use of underground facilities and many more characteristics of constructed facilities are influenced by climate and geology.

Local construction industries are very competent at producing buildings and infrastructure which deal with their normal conditions. When climate or geology behaves in unpredictable or extreme ways, construction can be caught out. Heavy snow or rain, unusual cold or heat, or high winds can all damage buildings and infrastructure or cause it to malfunction. Earthquake zones pose particular problems for construction. Modern design is able to cope competently with earthquakes although the necessary safety provisions are expensive. However, much of the built environment was constructed before these modern technologies existed and earthquakes can cause old buildings and infrastructure to collapse. The results are devastating; people are killed and injured, television cameras move in and in many cases the country affected has to appeal for international help and aid.

Other combinations of climate and geology can cause freak conditions which overwhelm buildings and infrastructure. Mud slides have devastating effects in many parts of the world and the incredible dangers posed by tsunamis was tragically demonstrated in coastal regions and islands around the Pacific Ocean in 2004; and again in Japan in 2011.

Floods are widely seen as an increasing problem. This is partly because many countries feel forced to build on flood pains as safer areas are already crowded. This combined with global warming which is widely expected to increase the incidence of serious floods raises massive problems. Novels looking ahead to 2020 and beyond describe scenarios where, for example, the United Sates needs to relocate as many as a hundred million people because of rising sea levels. Other parts of the world face greater and more imminent threats from floods. Construction has a major role to play in providing defences against rising water levels and designing and constructing buildings and infrastructure able to withstand the effects. It is at least arguable that construction managers should council against new construction where there are serious risks of flooding.

2.4 Economy

Economic conditions have major impacts on construction. When economies are growing, people feel confident so they spend money easily and are prepared to invest in the future. The demand for new construction increases. Resources become harder to find, costs rise but prices for new buildings and infrastructure rise even faster, so construction companies can make profits. These conditions are reversed when economies stagnate or decline. Confidence is low and the demand for construction reduces. New construction depends increasingly on public sector investment. Construction companies fail and only the most efficient survive and then only on minimal profits.

In considering economic conditions it is important to recognise that national economies are interlinked and many activities are shaped by the global economy. The financial crisis of 2007 began in the United States but spread rapidly and within less than two years had affected every developed country. These events provided many lessons about the way economic booms feed on themselves and perhaps the inevitability of periods of rapid growth being followed by downturns. They showed the fragility of financial and economic models and perhaps the main lesson was that economic forecasts need to be treated with great caution.

Construction is inevitably caught up in these fundamental uncertainties. Orders for new construction are highly dependent on economic conditions and peoples' expectations about the immediate future. When people feel confident about the future, they will invest in high-quality buildings and support major improvements to the infrastructure. In different economic conditions, when short-term problems and dangers dominate thinking, people are reluctant to sanction new construction and any that does go ahead has to be produced efficiently at low costs. Beyond these immediate issues, construction has to work on the basis that in the long term there are few certainties. Construction managers need to take these economic facts of life into account in deciding strategies for their companies and projects.

2.5 Government

Government regulates the construction of the built environment in the public interest. Some governments are very restrictive while others are willing to allow almost anything to be constructed. In deciding what to regulate, national and regional governments are influenced by the existing level of development, the land available for various uses, economic circumstances and their own priorities.

Policy objectives change over time, often in response to a failure of one regime to achieve what was intended. At various times national and regional regulators have tried to determine population densities, the location of industry, the provision of public spaces, the economic or even the physical health of residents, the provision of low-cost housing, the appearance of buildings, the availability of public transport and the kind of restaurants which can be established in a wealthy district. Some city planners produce master plans which are very detailed in specifying what may be constructed. Others define zones for commercial or residential development backed up by general rules about the size and nature of buildings which will be allowed.

In practice it is difficult to ensure the particular outcomes desired by government. Planning authorities propose many things but it is for construction customers and companies to put them into effect. They are influenced by many forces in addition to government regulation and the most experienced and successful construction customers are wonderfully creative in finding ways to use planning regulations to their own advantage. Equally, some planning authorities are creative in granting permission to develop a particular site in exchange for what is called planning gain. This may require the development to include public spaces, low-cost housing, sports halls, road improvements or almost anything else considered desirable in the public interest. The overall results inevitably are mixed. Planning regulation has preserved important parts of the built environment and open countryside. Equally it has distorted much development leading to the over provision of some kinds of buildings and serious shortages of others.

Governments also regulate the design and construction details of individual buildings and infrastructure. The aim is to ensure the safety, health, welfare and convenience of people in and around new facilities. These regulations provide detailed rules which in many countries now cover such matters as water and energy conservation, insulation, access for disabled people, the use, storage and disposal of toxic substances, fire safety, ventilation and air quality and other politically sensitive problems.

2.6 Culture and fashion

Construction is influenced by national cultures and current fashions. In some countries it is normal to use new facilities as a public display of wealth or power. This is regarded as vulgar in other nations and people commissioning new construction expect the resulting facilities to respect the existing built environment and fit in unobtrusively. National cultures influence the appearance, particularly of buildings. There tends to be a consensus on how houses, shops, offices, schools, government buildings and all the other types of building should look. Where similar views persist for decades or even centuries, fashion has little influence. Particularly in towns and cities which have a long history and are widely regarded as beautiful, established cultural values tend to determine what is built. This also is the case for new construction in areas of outstanding natural beauty where people want buildings to blend in with the countryside.

Current fashions in design and architecture are more evident in towns and cities shaped by dynamic change. The world's first skyscraper was completed in Chicago in 1885 as the city was establishing itself as the major transport and telecommunications hub linking the eastern and western states of America. The Home Insurance Building was designed by William Le Baron Jenney, an engineer (Figure 2.5). It was ten storeys high and it provided the first use of structural steel although the building's frame mainly comprised cast and wrought iron. The key development in making such a high building feasible was the development of safe elevators by Elisha Otis.

In the twentieth century Le Corbusier's pioneering ideas for urban living, which he first expressed in the 1920s, were not realised until the 1940s when they were seen as one answer to the chronic housing shortages caused by World War II. The prime example is the twelve story apartment block, Unité d'Habitation in Marseilles shown in Figure 2.6. It houses 1,600 people in a rectilinear concrete grid which accommodates a variety of individual apartments to provide an outstanding solution to crowded urban living. It started a major new fashion for high-rise apartments which produced some outstanding architecture. Tragically Le Corbusier's ideas were copied by lesser architects producing increasingly high and depressingly sterile apartment blocks. The results proved to be urban misery for millions of people, a reaction against high-rise apartments and a new fashion for low-rise housing.

The United Kingdom's financial deregulation in 1986 resulted in New York and Tokyo style financial centres becoming fashionable in the solidly traditional

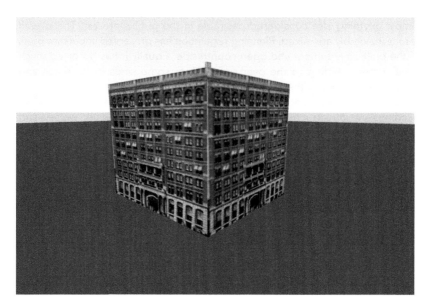

Figure 2.5 Computer Model of the Home Insurance Building.

City of London. Broadgate, in part constructed in the air space over Liverpool Street railway station, anticipated this change. It provides a group of impressive high rise buildings designed to accommodate the financial trading rooms made possible by deregulation. It also provides new open spaces which give the 30,000 people who work there each day places to relax, be entertained, get a meal or go shopping. The whole development adds a stylish new district centre to the City of London. An even more dramatic example of the new fashion is

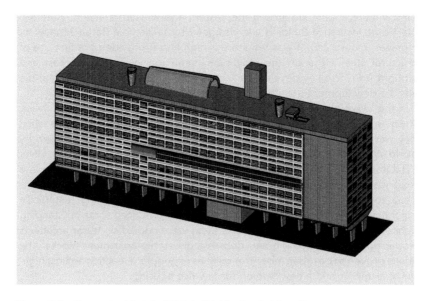

Figure 2.6 Computer Model of Unité d'Habitation in Marseille.

Figure 2.7 Computer Model of the Canary Warf Financial District in London.

Canary Wharf, shown in Figure 2.7, which provides a serious rival to London's traditional financial centre. This purpose built district on the edge of the old city provides a stylish group of skyscrapers. It includes, at the time it was constructed, the United Kingdom's tallest building and in total provides offices for over 100,000 people employed by banks and other financial institutions.

Construction managers need to be aware of the powerful forces exerted by culture and fashion in shaping what may and may not be constructed. They should remind customers and designers about these sometime intangible constraints; and ensure they take account of the very real constraints available to local planners, politicians and special interest groups if new construction threatens to offend established cultural values or currently fashionable ideas about the kind of buildings or infrastructure which should be allowed.

2.7 Technology

Construction is directly shaped by available technologies and this is self evident simply by looking at buildings and infrastructure. Construction managers need to understand established technologies because they affect many of their decisions. More than this they need to be aware of likely new developments. Partly this is to be ready to use new technologies when they arrive but also to influence research and development so it produces technologies which will be of the greatest value to construction.

The internet now makes information available where and when it is needed. It allows design offices, factories and construction sites to be linked to form integrated construction systems. The same technologies that drive the internet enable buildings and infrastructure to behave intelligently. Combined with materials able to respond to changing local environments, information technology has the potential to give people fine control over the built environment.

The drive for sustainability is already providing new technologies which enable buildings and infrastructure to generate power, conserve water, treat waste and have a minimal impact on the environment. Construction managers should be involved in ensuring these and other future developments meet the needs of the built environment and their construction companies are prepared to use the full potential of new technologies.

2.8 Customers

The primary interface between construction and other human activities is provided by customers. These are organizations with the power or wealth to initiate construction projects or buy a new facility after it has been constructed. To varying degrees all customers help shape what, where and how construction is undertaken. Customers able to command substantial resources invest in the built environment to meet specific needs but also to express their wealth, power, culture, style and beliefs. The results can be truly remarkable where rich and powerful people have influenced a city for thousands of years. This is wonderfully evident in the ancient cities of Jerusalem, Rome and Beijing. Their complex histories, driven by ever changing interactions between power, defence and religion, have produced built environments that inspire and challenge every visitor.

Most construction customers have less influence than the powerful individuals and governments who shaped these great cities. Whatever their needs and resources, construction managers need to take account of customers' needs and the resources they command and where necessary guide them into more informed decisions. The next sections provide guidance on this important task by describing the influence of some distinct categories of customers and the resulting manmade facilities that form the built environment.

Government

In many regions and countries government bodies are the largest and most influential customer for construction. They need buildings for most of the tasks they undertake (Figure 2.8) and often directly provide at least part of the national infrastructure. This gives them the power to influence every aspect of construction. The nature of their demands, the way they define their requirements, the way they select construction companies, the kind of contracts they are prepared to agree, and the way they treat the construction companies they employ affect the built environment and the performance of the construction industry.

Governments can behave in ways which would be virtually impossible for other customers. La Défense in Paris was created because three imperial Presidents of the French Republic, General de Gaulle, Giscard d'Estaing and Francois Mitterand, wanted a major commercial centre in their capital city. It is now the largest purpose-built business district in Europe providing over 3.5 million square metres of office space which serves at least 150,000 workers. The President's individual power was illustrated by Francois Mitterand in August 1983 when he resolved a long running problem by deciding to build the Grande Arche shown in Figure 2.9. It provides a focus for the western end of

Figure 2.8 Computer Model of the London City Hall.

Paris' historical axis which runs from the Louvre, through the Arc de Triomphe to the Grande Arche.

The decision to construct this important building was made by President Mitterand himself. Before agreeing to the construction of the 30 storey high, open cube, he demanded a demonstration of its visual impact. This was achieved by positioning the tallest crane in France on the six-lane motorway which loops around La Défense. The crane carried a massive reinforced concrete slab, 20 metres long and painted white to look like marble. The slab

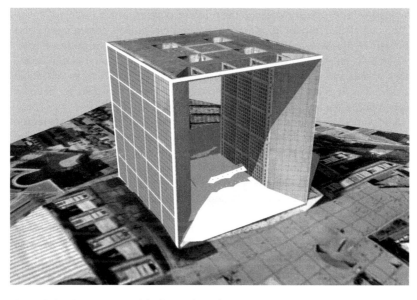

Figure 2.9 Computer Model of Grande Arche in Paris.

was positioned by teams of engineers 110 metres above the ground exactly where the top of the Grande Arche now sits. The incredible task of bringing the crane and its load into Paris and getting it into position required a meticulously planned, week long journey which stopped the traffic over large parts of Paris. After spending two days looking at the effect of the gleaming white slab on many of the most historic views of Paris, President Mitterand ordered the building of his great monument.

It is not only new districts which are planned in this way. Whole cities result from individual decisions. These include major capital cities such as Beijing, ordered by Kublai Khan in 1267; Washington DC founded in 1790 on the basis of a compromise over the national debt on a site selected by President Washington; Canberra, officially designated in 1911 was a compromise between the competing claims of Sydney and Melbourne; and Brasilia, inaugurated in 1960, was designed in part to reduce the impact of popular revolts by moving the government away from Rio de Janeiro and most of the population.

Housing

The demand for housing gives rise to more buildings than any other single force. People need somewhere to live. The majority of people, and in developed nations this is an overwhelming majority, live in housing which is reasonably adequate. Many of these people buy a house or apartment. This is for most people by far their largest single financial decision and their house or apartment is their most valuable possession. Sophisticated financial arrangements have developed to make the purchase possible for people who otherwise would never have sufficient money. Governments actively aim to ensure that as many people as want, can buy somewhere to live.

Renting has some advantages. It provides flexibility since it is easier to move to a different location if this does not depend on selling a house or apartment. Normally it is initially cheaper to rent than buy and this makes housing available to people with less money. It also has disadvantages since people are less likely to maintain or improve dwellings they do not own.

The overall demand for housing and the nature of what is constructed is also affected by the average number of people per dwelling. Societies characterised by large family groups require far fewer dwellings than societies where many people live alone. In many developed countries, the number of people per dwelling is reducing due to an increased incidence of divorce, children leaving home earlier and people living longer.

In addition to whether people buy or rent their dwellings and the average number of people per dwelling, there are other forces which shape the kind of housing produced in individual cities, towns, villages and rural locations. These include climatic and geological conditions, economic realities, government policy, culture and fashion, the available technologies and individual wealth. The results may be huge apartment blocks crowded together to provide social housing; elegant buildings in beautiful surroundings which provide luxurious apartments for wealthy individuals; street after street of virtually identical low rise houses; large, individually designed houses surrounded by manicured gardens accessed along tree lined lanes; or great mansions surrounded by huge estates.

Outside the formal built environment, many major cities are blighted by slums in informal collections of rudimentary shelters. These represent a massive failure of the built environment to provide sufficient housing. Meeting the obvious need for more good housing should be a universal responsibility and is certainly a matter for construction managers to think about.

Religion

Great religious buildings around the world reflect many different beliefs as well as distinct cultures, histories and architectural styles. Saint Peter's Basilica in Rome clearly embodies Christian faith but it also tells us about the power and influence of the Catholic Church, provides wonderful examples of Renaissance architecture, sculpture and painting and is a major tourist attraction. Similar factors influence all great religious buildings. There are more modest churches, mosques and temples in every city, town and village. They are often important landmarks and provide a focus for many important activities.

Religious or other deeply rooted belief systems can influence the built environment in other less obvious ways. They may directly influence the kind of buildings which are constructed. Many schools, hospitals, care homes, shops and other buildings are influenced by the users' religious beliefs.

Manufacturing

Manufacturing companies have been significant customers of the construction industry for at least 200 years. Their demands have changed both in the terms of the nature of the buildings and infrastructure they need and its location. The industrial revolution began in Britain with the mechanisation of textile industries, the development of iron-making and a great increase in the use of refined coal. Canals and railways enabled large factories to be located near basic resources and their products to be widely distributed. Steam power then increased production dramatically and machine tools facilitated the production of an ever increasing range of machines for different industries.

Industrialisation rapidly spread to Europe, North America and Japan. The internal combustion engine and electrical power generation dramatically increased industrial output and the ease with which factories could be established in new locations. Manufacturing is now a part of virtually every national economy.

Efficient manufacturing requires thousands of workers and their families to be housed near factories. As a result, early industrialisation caused many towns and cities to grow rapidly. Some provided terrible living conditions, forcing factory workers into small houses which lined cramped streets. Others were a model of urban development. New factories today often provide the centre of new residential districts. This is particularly the case in those Asian countries where much traditional manufacturing is now undertaken. The purpose built industrial districts provide greatly improved living conditions for many workers who have moved from poor rural communities.

In the developed world today manufacturing tends to be concentrated on high technology products. Modern factories look more like high quality offices. The best are purpose built and often form a key part of elegant, new residential districts. Others are clustered together on well-designed industrial estates. At the opposite extreme, in poor countries or cities, much basic manufacturing takes place in old buildings which provide inadequate and dangerous working conditions.

Manufacturing has always provided distinct challenges for construction customers and indeed for everyone involved in decisions about the built environment. They provide work which is essential for many people but many industries are dirty, noisy and produce considerable pollution. Modern industries are cleaner and less obtrusive but they increase road and rail traffic. All these factors mean decisions about the location of factories are always controversial.

There is another very different challenge which results from changes in manufacturing. Factories become obsolete as new technologies are developed and lower cost competitors emerge. Old industrial estates become uneconomic and are abandoned. Many such sites are badly polluted by industrial waste and are rarely in prime development locations. Finding new uses for the buildings or demolishing and replacing them needs considerable financial investment and provides major challenges for everyone involved.

Commerce

Commerce directly influences what is constructed and where it is likely to be located. Commerce is defined as the buying and selling of goods and services. It takes many forms and the internet is adding to these at a startling rate.

The great financial centres of London, New York and Tokyo include commercial districts which were able to grow at a startling rate by exploiting the power of market forces (Figure 2.10). These are clearly marked out by huge skyscrapers clad in glass and stone. These massive buildings house the global systems which determine the wealth of nations. The buildings were designed to be reassuring symbols of financial strength but when the 2007 financial crisis damaged so many lives around the world, these great monuments to capitalism became the focus of violent protests and hatred.

Figure 2.10 Computer Model of the World Financial Centre in New York.

Shopping is the other type of commerce which has a very direct impact on the built environment. Throughout history there have been shopping areas at the centre of urban areas. Traditionally shops defined the main street but increasingly they are grouped into purpose-built shopping centres. Some are located in urban centres, usually fronting onto the main street but they are also served by convenient access from dedicated car parks. Increasing numbers of shopping centres are located on the edges of urban areas because this provides easier access for customers and their cars. The internet is affecting shopping to an ever increasing extent as people discover the convenience of shopping online. This is creating a demand for a network of huge and ever more sophisticated distribution warehouses with easy access to motorways.

Education

Schools and universities require various kinds of buildings and infrastructure. Many educational buildings provide straight forward teaching spaces. Others, especially in universities, have distinct and often very specific requirements because of the specialised nature of the teaching or research they will house. Other buildings are needed for various support functions including offices for academics and administrators, dining rooms, sports halls, swimming pools, halls of residence and various security provisions. The infrastructure may be required to provide car parking, sports pitches, playgrounds and peaceful places to sit and think. It may also be arranged to deter intruders.

The nature of educational buildings is greatly influenced by the money available to construct them. Some educational establishments have considerable financial resources and expect new buildings to provide a public display of their wealth and at the same time represent the high-quality education they provide. Other schools and universities have less money and need to ensure they get the maximum amount of teaching space for their limited budgets. The source of the funds also influences what is constructed. Much education is funded by government who normally impose stringent rules which govern the form and style of educational buildings. There is often greater design freedom for privately funded educational buildings and this can be seen in many elegant schools, colleges and universities around the world.

Health care

Modern hospitals provide what may be the most challenging kind of building to construct. Operating theatres require a bewildering range of services to be reliably available in sterile spaces. Wards need to provide safe, comfortable spaces for patients yet support difficult and potentially dangerous work by nurses and doctors. Casualty departments need ready access for people, ambulances and helicopters around the clock; they need access to a wide range of diagnostic equipment; must be ready to deal with critical medical conditions; and cope with a great variety of people some of whom may be mentally ill and pose real threats to themselves and others. Hospitals deal with out-patients as well as in-patients. They need examination rooms, waiting rooms, toilets, restaurants and cafes. Other types of specialised departments range from maternity to mortuary.

This is all challenging but it can be further complicated by new developments in medical science requiring substantial changes to the kind of spaces needed.

Equally a doctor newly appointment as head of a major department may want to introduce highly individual ideas about the spaces he needs to work in.

Health care means more than major hospitals. It requires general practitioners' surgeries, health centres, clinics, dentists' surgeries, opticians' consulting rooms and a variety of spaces used by the practitioners of alternative medicine.

Transport

Transport systems provide the most extensive elements of the built environment. Railways circle the globe many times and enable passengers and goods to be carried to every major city and town. Railway lines need tunnels, cuttings, bridges, viaducts and embankments. Railway systems need a range of engines, passenger carriages and goods trucks. Stations range from simple halts to city centre buildings which incorporate a variety of sometimes beautiful spaces. Beyond this railways require sophisticated and reliable signally systems, large goods yards and sidings, and maintenance and repair shops.

Modern road systems are even more extensive. They link motorways to major roads, city streets and country lanes. They have systems to control the flow of traffic; systems to deal with accidents; and are supported by filling stations, places for rest and refreshment, and extensive car and lorry parks.

Airports are needed to enable millions of people to fly to wherever they need to be. They also facilitate the rapid transport of high-value goods over long distances. Docks enable bulk goods and containers to be transported around the world economically. Increasingly they support the activities of massive cruise liners. Harbours and marinas provide small boats and yachts with safe berths when they are not at sea. In total, transport systems form a vital part of modern life.

Energy and basic resources

The provision of power and energy is indispensable in the modern world which is ever more dependent on sophisticated machinery. A diversity of power plants convert the energy of moving water and wind, sun, nuclear fuel cells, heat created by burning fossil fuels and so on into electricity. It is then regulated by transformers and transported over long-distance transmission systems to households, factories, office buildings and all the other facilities which require electrical power. Much of the energy infrastructure is concentrated in urbanised areas but it also extends to rural and unpopulated areas because the sources of energy themselves are often in remote locations.

Oil and gas are extracted from earth's crust on- and off-shore through large wells and transported through vast networks of pipelines to be processed in refineries. It is then stored in huge tanks ready for distribution to petrol stations, households, factories and other facilities.

Coal and other materials are extracted in surface and underground mines all over the world. They are often processed on-site before being transported to storage facilities and processing plants for further treatment.

Modern life depends on vast quantities of many kinds of extracted raw materials and energy being distributed to every part of the built environment. As a consequence, modern landscapes are dotted with the facilities needed to provide energy and basic resources; and are crisscrossed by complex infrastructures comprising power lines and pipelines either above or below ground.

Waste treatment

Human activities produce a huge amount of physical waste that needs to be treated before being released back into the environment. From simple house-hold wastewater to toxic manufacturing by-products, almost every manmade waste needs to be processed in waste treatment facilities many of which are extremely large and complex.

For instance, much wastewater is purified in mega wastewater treatment plants fed by incredibly complex and extensive wastewater infrastructures. Solid waste has traditionally been disposed of in vast landfill sites but recently such waste disposal is increasingly unattractive due to a lack of appropriate sites and hazardous local impacts. Recycling and reuse facilities are becoming more and more common throughout the modern world.

Highly toxic industrial waste often requires specialised storage and waste treatment facilities. In some cases these facilities form part of large industrial complexes, in others they are separate and may serve a single or several industrial zones. A growing problem is dangerous waste which cannot be treated. For example, the only viable option for disposing of nuclear waste appears to be underground storage which needs to remain secure for centuries.

Water supply

Water is one of the essential requirements for life. There would be no life on Earth without this precious resource. The ability to survive in sometimes very harsh conditions depends on the availability of water. That is why water infrastructure, along with roads, shelter and security, forms humankind's very first substantial constructed facilities. Water is essential for drinking, pro-ducing food, most manufacturing processes, heating systems, swimming pools and many more elements of modern life. All this depends on a vast water supply network to enable people in the developed world live their normal lives.

Ancient civilisations built aqueducts and drilled simple wells, some of which remain in use even today. The modern demand is such that humans drill extremely deep wells, build reservoirs and water supply pipelines, all in the search for fresh, clean water. In many parts of the world, water in drinkable condition is an extremely scarce resource. Processing facilities clean and purify whatever fresh water is available before it is distributed for drinking. In increasingly large areas of the world, processing has to be taken further and desalination plants provide the only viable means of providing drinking water.

Communication

Technological developments are creating ever more sophisticated solutions that make communication over great distances straightforward and efficient. The advent of telecommunication channels of various kinds has given rise to vast telecommunication networks which, like the energy infrastructure, extend far beyond urban areas. Physical networks of various types have sprung up throughout the world, which in turn means there are now thousands of miles of transoceanic cables crossing even the largest and deepest of oceans.

The recent explosion of wireless communications requires an increasingly complex network of transmitter masts, central antenna hubs, transmitting stations, all interconnected by hundreds of thousands of miles of cables.

The past few decades have seen the beginning of satellite telecommunications which has extended the built environment into outer space. According to NASA there are currently around 3,000 artificial satellites orbiting the earth. Even more manmade objects can now be found throughout our solar system and our space activity is set to increase exponentially over the next few decades. We are witnessing the birth of outer space construction which may well affect the way we construct facilities on Earth.

2.9 Buildings and infrastructure

The preceding sections described the forces which shape the complex built environment illustrated in Figure 2.11. The description identified some differences between buildings and infrastructure but the distinctions have important consequences for construction management and need to be considered further.

Architects design buildings while civil engineers design the infrastructure. This divide is not entirely precise. Civil engineers are often involved in designing parts of buildings and architects are often brought into the design teams responsible for additions to the infrastructure. This overlapping of responsibilities does not blur sharp differences between the nature of architecture and civil engineering.

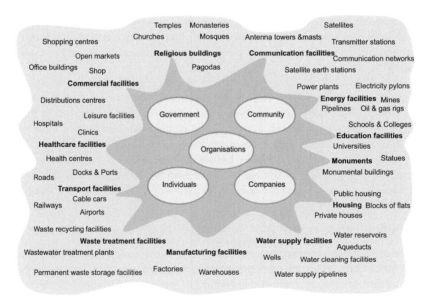

Figure 2.11 Cross-Section of the Built Environment Needed for Modern Life.

2.10 Architecture

People tend to have a more direct and longer interaction with buildings than with infrastructure. This means greater care is justified in controlling the way they affect our senses. People expect buildings to provide delight and pleasure beyond that provided by their basic functions. We need houses to provide shelter, safety and convenience, schools to support learning, hospitals to make us healthier, theatres to provide places where we can be entertained in comfort, and prisons to keep dangerous criminals off the streets. Beyond these basic functions we expect buildings to be beautiful, exciting, challenging and in every way significant. In other words we view them as works of art.

Architecture is a particularly challenging art because buildings involve many distinct technologies. Each individual technology is supported by design, manufacturing and production methods which to a greater or lesser extent are tried and tested. The particular problem posed by buildings is the relative large numbers of junctions between technologies. These need to be produced taking account of the specific conditions in individual buildings.

Some junctions between distinct technologies are well developed in established local practice. Traditionally junctions between brick walls and wooden window frames were a matter of established craft practice. As more new and very different technologies were introduced into modern buildings, craft practice was unable to cope. Buildings had to be designed in detail and it is the prime responsibility of architects to produce coherent, overall designs for buildings.

Architectural design is undertaken in markedly different ways in different countries. This is illustrated by the three influential approaches described in Chapter 1. Architects may concentrate on the overall appearance of their buildings as in the American approach. They may insist on designing every detail to ensure the total integrity of their buildings as in the British approach. They may be part of an integrated team responsible for all the activities involved in the construction of buildings as in the Japanese approach.

The common driving force in architects' work is the production of architecture rather than mere building. It is concerned with the overall style, shape and appearance of buildings as well as their function. Architecture provides cultural symbols which people imbue with meanings. It makes people stop and look. It forces them to make judgements. The difference between architecture and civil engineering is a matter of art.

2.11 Civil engineering

The need for a developed infrastructure emerged towards the end of the eighteenth century as large cities became unbearable for all except the richest people. Terrible housing, poverty, disease, infant mortality, crime and degrading work could no longer be tolerated. Sewers and water supply systems became absolute necessities. Some European cities cleared large areas of their overcrowded centres to form wide boulevards to provide routes for sewers and water supply lines. Paris and Barcelona are two of the best examples of the dramatic improvements in living standards which resulted from this government driven approach. London relied more on private enterprise so sewers and water supply lines had to be threaded through existing street patterns. There were a few bold initiatives, for example, the Thames Embankment created a wide road and housed a massive sewer.

The knowledge and skills needed to design and construct this first basic infrastructure were called civil engineering to distinguish them from military engineering. The development of the railways made further demands on civil engineers and in 1818 the Institution of Civil Engineers was founded in London and the eminent engineer, Thomas Telford, became its first president. The profession deals with the design, construction and maintenance of infrastructure projects including bridges, roads, canals, dams, railways, tunnels and flood control measures.

Compared to buildings, infrastructure involves relatively few distinct technologies. However, new infrastructure tends to be large and technically difficult and have to be constructed in difficult conditions. It may cut through densely inhabited areas, involve excavating or tunnelling into difficult geological conditions, or face the threat of floods.

The large scale of much new infrastructure combined with the high degree of technical and environmental difficulty defines the main challenges faced by civil engineers. First, they have to solve difficult engineering problems and then ensure their designs are properly and safely realised on large, often fragmented construction sites. This requires responsible civil engineering involvement at all stages of their projects. This is achieved by having designs produced by civil engineering consultants and direct construction work undertaken by civil engineering contractors. The common professional background provides many advantages, not least in providing unusually coherent project organizations. More importantly it ensures the competence and safety of important elements of the built environment.

2.12 Thinking about the forces which shape the built environment

This chapter has described the main forces which shape the built environment. Their nature and impact have changed over the centuries and they continue to pose new challenges for construction managers. A useful way of thinking about these forces is to identify the changes already taking place and speculate about which are likely to become significant in the next few decades.

Exercise

Imagine that you run a housing development company that aims to develop an estate of 100 low energy houses of the type shown in Figure 2.12 on a brownfield site. Have a close look at Figure 2.11 and think how such development would alter the existing environment. Specifically, what elements of the built environment would you need to add? Make a list of

Figure 2.12 An Image Showing One of the 100 Low-Energy Houses that You are to Develop.

elements and then think how you could make each of the new elements with the minimum possible environmental impact.

Hint: Think about elements like the houses themselves, utilities, new landscaping, streets and roads, street furniture, etc. What utilities could you run underground? Could you build houses to be partially or completely self-sufficient (e.g. wastewater treatment plants instead of sewage; local renewable energy sources as opposed to connecting houses to the electrical grid; etc.)? What difficulties would you face in trying to minimise environmental impact through the above methods (e.g. legislation, low sun index affecting the use of photovoltaic systems, access to public transport, etc.)?

Answer the following questions:

1. What forces shaped the nature, size and appearance of the development?
2. What organizations were involved in determining what was constructed?
3. What would have been needed to make the development significantly better?

It is also worth considering how the forces which shape the built environment could be mobilised to help solve major problems. These include:

1. Providing decent housing for everyone rather than the slums which blight many major cities.
2. Providing efficient, affordable and non-polluting transport systems.
3. Avoiding new developments on land subject to flooding.
4. Avoiding new developments on land needed for agriculture to feed a growing population.
5. Ensuring the built environment does not generate greenhouses gases.
6. Providing properly rewarded employment for everyone who wants to work.

Further Reading

The following publications are the source of ideas used in this chapter and provide further information for readers.

Brandon, P. S. and Lombardi, P. (2010) *Evaluating Sustainable Development: In the Built Environment*. Wiley-Blackwell, 2nd edn. The book proposes a framework for evaluating sustainable development of the built environment. The fifteen modal aspects and their role in the context of sustainable development in the built environment are particularly of interest and three case studies show how the proposed multi-modal framework can be used as a decision-making tool.

Halliday, S. (2007) *Sustainable Construction*. Butterworth-Heinemann. This book provides a good overview of the subject with well thought through and extensively illustrated examples covering social as well as technical aspects. The case studies are succinct and well supported. The chapters begin with clear introductions written in an easily accessible form using text inserts and a clever selection of visuals in various formats (e.g. images, diagrams and tables).

Sudjic, D. (1993) *The 100 Mile City*. Harper Collins. This book describes important examples of the forces shaping modern cities with a particular focus on London, Paris, New York, Tokyo and Los Angeles. In essence, the book is a multi-faceted critique of the modern development of cities beyond their former limits resulting in a decay of their inner core. It provides a sobering insight into costly development-driven expansion.

Chapter Three
Construction Concepts

3.1 Introduction

The chapter identifies and defines the concepts needed to understand construction management. It is an essential introduction to Chapter 4 which provides the theoretical basis for construction managers to make construction efficient. The resulting theory is used in Chapters 5 to 9 to explain each of the distinct approaches to construction management currently used in the world's major construction industries. In this way the chapter provides an essential first step towards empowering everyone involved in construction to make decisions which improve their efficiency and the quality of construction products.

It is central to achieving these high objectives to recognise that projects need to be well managed and companies need to be equally well managed. The book is based on the central assumption that construction management is at the centre of both company and project management; and this integrated understanding is vital for the construction industry'sl prospects.

3.2 Construction products

The first basic concept needed to understand construction management is the **built environment** which, as Chapter 2 describes, comprises all currently existing constructed facilities. A **facility** is a permanent, fixed product constructed by human organizations which alters the environment. The facility may alter the existing built environment or create a new built environment hub in a virgin natural environment.

This definition glosses over a variety of structures which may or may not qualify as part of the built environment. Tented villages which may remain on the same site for months are one example of facilities of ambiguous status in terms of the built environment. Fortunately this uncertainty does not influence the theory of construction management and the vast majority of facilities clearly and unambiguously form part of the built environment.

3.3 Customers

Construction actions are initiated by a customer. In addition to deciding they need a new facility, a key part of the customer's role is ensuring construction actions are financed. Many construction professionals refer to their customers

Construction Management Strategies: A Theory of Construction Management,
First Edition. Milan Radosavljevic and John Bennett.
© 2012 John Wiley & Sons, Ltd. Published 2012 by John Wiley & Sons, Ltd.

as clients. This terminology does not add anything to the concept and so the wider term of customer is used throughout this book.

A customer is normally the owner of the new facility produced by the construction actions they initiate. An **owner** is an individual or organization which is the legal possessor of a constructed facility. However, the customer may be an organization which has a relationship with the owner. They may, for example, be an organization which will lease the facility when it is complete and have agreed to act as the customer for the construction actions needed to produce the new facility. So irrespective of their interest in the new facility, a **customer** is an individual or organization external to the construction industry which initiates construction actions. The actions normally produce a new facility but they may be applied to an existing facility to alter or extend it. Throughout the book, the term new facility includes entirely new facilities, and alterations and extensions to existing facilities.

3.4 Construction actions

There are many practical guides to the set of actions needed to produce a new facility. In various terms they all include seven essential actions.

The first essential action is preparing a **brief**. A brief is a statement describing a new facility required by a customer and identifying constraints which must be observed in undertaking the necessary construction actions. It normally states the customer's objectives, and describes the functionality and performance required in the new facility.

A brief results from actions by a customer which establish the need for a new facility. It should be prepared carefully to ensure the completed document describes everything the new facility must provide. The brief should provide all the information needed to allow a complete design to be produced. This has a direct influence on how efficiently the new facility can be produced because inadequate briefing almost inevitably results in late design changes which interrupt and delay the subsequent construction actions.

Preparing a competent brief usually involves extensive consultation and discussion within the customer's organization. These internal actions may identify the need for expert help in evaluating possible sites for the facility, the financial, legal and bureaucratic hurdles to be negotiated and the way the new facility will be used.

Briefing is often considered to constitute the first stage of a project. However, customer requirements do not arise suddenly. In most cases a new facility is required due to operational constraints which have developed over time. For example, a new hospital building is needed because existing hospitals in the area cannot cope with an increase in the number of patients due to population growth, or a new office block is needed because currently available facilities in the area are either full or inadequate. There may be a need for better facilities to accommodate modern processes or more people. Needs usually develop over time and customers often initiate a project when the build up of constraints is such that they start affecting operational efficiency and effectiveness. However, the need is not always as rational as this and some customers order new facilities for prestige and aesthetic reasons with functionality being of only secondary importance and so is considered during the later stages of preparing a brief.

Operational requirements can be expressed in the form of building capacity, production capacity, technology compatibility, building performance, and so forth. Long-term observation and recording of operational efficiency and effectiveness provides a robust basis for producing effective briefs. The recording should be systematic and ideally spans several years to provide a comprehensive operational constraints report which then informs the briefing process. The more complex and capital-intensive the facility the greater the need for a comprehensive operational constraints report.

The operational constraints report should identify major obstacles to efficient and effective operation which make clear the need for a new facility. For example, current factory facilities may offer sufficient space for the existing plant and equipment but be inadequate for a new production line which is bigger or requires more sophisticated engineering services and so needs a different kind of facility.

Example – Operational Constraints Record

Consider an existing hospital surgical tower where records of operational efficiency and effectiveness highlight the following constraints:

- Operating theatres deemed too small due to increasing use of high-tech surgical and auxiliary equipment (based on surgeons' views of the optimum size and shape of operating theatres; and records showing equipment has to be stored at different locations and brought into the operating theatres as and when it is needed causing delays).
- Staff rooms considered too distant from the operating theatres which wastes staff time due to the need to constantly monitor the equipment (based on nurses and doctors' views of the optimum distance between staff rooms and operating theatres).
- Energy consumption disproportionally high due to poor insulation and uneven distribution of heat throughout the building (based on temperature monitoring which shows some rooms and theatres to be significantly cooler or warmer than others).
- Insufficient number of operating theatres due to increased numbers of patients (based on records of the past 10 years which show a 10 per cent annual increase in the number of patients; and records of the number of occasions when waiting times have put patients at risk).
- Hospital wards too distant from the operating theatres (based on doctors and nurses' reports that the time taken to move patients to and from an operating theatre was unacceptably long).
- There are insufficient elevators in the building (based on observations of the elevators in use which show visitors often create congestion because they use the special hospital bed elevators when the passenger elevators are busy).

Think how the constraints identified in the operational constraints report could inform the brief for a new facility to replace the existing surgical tower. What else could you take into account which may inform the brief and benefit the design?

The brief is often recorded in a document describing in detail what the new facility must provide and listing the financial, safety, time and other constraints which must be observed in undertaking the necessary construction actions; but it can be a short statement describing the new facility by reference to some established norm or an existing facility. Various, more or less effective briefing processes exist in practice. The theoretical implications of the most important, including their influence on subsequent construction actions, are described in more detail in the Chapters 5 to 9.

The second essential action is preparing a **design**. A design is a statement describing the product of construction actions which satisfy the requirements of the brief for a new facility. It remains common for designs to be recorded in paper based drawings and specifications. However, information technology enables designs to be recorded in various electronic forms. These include interactive three dimensional representations integrated with building information models which help evaluate designs. They can identify the resources required and calculate the likely times and costs of the subsequent construction actions. They can highlight potential problems and help evaluate risks associated with particular designs.

Increasingly sophisticated design technology is needed because many modern constructed facilities comprise many hundreds of different technologies. As a consequence the design process is often long and complex involving alternative design ideas from various sources being produced, considered, redesigned and reconsidered several times before final decisions are agreed. All this is further complicated by the need to take account of challenging legal and bureaucratic processes set up to protect the public interest in the built environment.

One significant result of the sheer complexity of the design process is that it is undertaken in distinct stages. This helps ensure the requirements set out in the brief have been understood and taken into account before significant effort is devoted to developing an inadequate design. The stages are given a variety of names which include concept design, scheme design, detail design, technical design and production information. Modern design management uses computer based tools which minimise wasteful iteration between design tasks. The best of these streamline the design process and provide a robust basis for aligning all the subsequent construction actions with the final design.

The third essential construction action is preparing a **plan**. A plan is a statement identifying the types of organization needed to produce a new facility and describing the constraints and targets which control the way these organizations need to work.

A plan may be little more than a statement of the budget and required completion date. More usually on construction projects of any size or significance, the plan includes a variety of detailed documents. These may begin with a project execution plan which provides detailed lists of the information needed at each distinct stage of the project. The plan for most construction projects includes a schedule of the time allocated to each action, and a financial budget and cost plan showing how the available money should be spent on the various elements of the project. The plan often includes provisions aimed at dealing with risk, quality assurance, safety, environmental protection and post-project evaluation. There may be an organization structure plan to establish how the organizations involved should relate to each other. There may be a schedule of regular meetings to ensure that decisions about the project are properly considered.

However simple or detailed, the plan should describe a framework which helps ensure design, plan and the subsequent actions of procurement, manufacturing, production and commissioning can be undertaken efficiently and safely, at times, costs and quality standards which satisfy the requirements of the brief.

The fourth essential action is **procurement**. This is the actions which ensure construction teams with the required skills, equipment and support are selected and employed; and arrangements are in place so the teams are supplied with the necessary materials and components and are motivated to undertake the construction actions needed to produce a new facility. Teams and resources have to be procured throughout construction projects usually beginning by appointing construction teams to produce the design and plan.

The term procurement is used in many of the practical guides to refer exclusively to the set of actions undertaken in a short, hectic period between the completion of the design and plan and the start of work in factories and on site. This is when companies are selected to provide construction teams to undertake the direct production of the new facility. Guides which take this narrow view tend to place considerable emphasis on competitive bidding, tough contracts and a multitude of provisions aimed at defending individual interests. This narrow view almost inevitably leads to inefficient performance.

Procurement needs to give careful consideration to the joint interests of everyone involved in producing new facilities if it is to establish a basis for efficient construction. This approach is intrinsic in the theory of construction management.

The fifth essential action is **manufacturing** which means the actions needed to make and provide all the materials and components needed for production. Traditional construction materials are manufactured as standard products and procured by project organizations as and when required. However, manufacturing plays an increasingly central role in modern construction. Large, sophisticated components manufactured in factories have radically altered production actions. Some of the best practical examples include operating theatres for modern hospitals, plant rooms which provide the core of heating and cooling systems for large commercial buildings, major elements of processing plants, and many more. The most sophisticated components incorporating complex service elements tend to be manufactured for a specific project.

Arranging for large prefabricated components to be manufactured and delivered to site exactly when they are needed is increasingly central to the success of construction projects. One important development is manufacturers developing ranges of standardised modules which can be incorporated in a variety of buildings. This innovation has the potential to transform the efficiency of the construction industry.

The sixth essential action is **production** which is the actions which convert materials and components into a new facility.

Production is normally called construction by practitioners but the term production is used for direct site construction to avoid confusion with the overall construction process. This terminology reflects an important change in direct site construction which is moving away from traditional crafts using standard materials. Modern construction projects rely more and more on the assembly of prefabricated components and so shares many characteristics of manufacturing. As described above, most prefabricated components are bespoke to meet the individual needs of a specific project but standard large-scale components are becoming available.

Production involves a number of stages each largely shaped by distinct technologies. Common stages in the production of a new building include: substructure, structure, external envelope, service cores, risers and main plant, entrance and vertical circulation spaces, internal divisions, and finishes. Other types of facilities involve somewhat different distinct stages in the production process which are well understood by specialists in the specific type of constructed facility. All these types of action share important characteristics which are fully taken into account by including production as the sixth essential construction action.

The seventh and final essential action is **commissioning**. This is the actions which turn the product of production into a fully tested and properly functioning new facility; and ensure the organization which will take over and run the new facility is trained to use and operate it.

Commissioning is often neglected as the organizations involved concentrate on completing production and handing over the new facility to the owner and moving on to another project. This is always a mistake because commissioning is a vital action which needs to be carried out professionally. It requires the individual systems and the new facility as a whole to be tested to ensure everything is fully working and satisfies the design criteria. It is equally important to ensure the organization which will take over and run the new facility is trained to use and operate it as intended and described in the brief.

Example – Commissioning and Facilities Management

Consider a situation where a school is being moved into a new building. Until now it was housed in a 150-year-old building with minimal services. Heating was the only centrally-controlled service in the old building. The new building includes a sophisticated Building Control System (BCS) which enables all major systems to be centrally operated through a modern user interface. The BCS comprises:

- a central security which allows all rooms to be locked from one central point;
- an automatic switch-off of all services in cases of emergency;
- energy management with centrally and automatically controlled humidity, air temperature, ventilation, air-conditioning, and electrical supply.

During the commissioning of the new building trade contractors thoroughly tested all the systems in the building in the presence of the school's facilities management team. Answer the following questions:

1. What is likely to be the single major obstacle to achieving effective use of the BCS?
2. Would it be possible for the new building to consume more energy than the old building given they are of equal size? Consider how this situation might arise.
3. How would you ensure the effective use of the BCS?

Construction organizations are involved in repairs and maintenance of facilities. It could be argued these constitute further construction actions. However, even minor repairs and maintenance work involves the seven essential actions. They may be small actions involving few resources but all seven actions are needed. In other words repair and maintenance is a sector of construction which does not involve essentially different actions from other sectors.

3.5 Construction

Construction means all human actions, including those undertaken with the help of or solely by plant and equipment, intended to produce and/or alter facilities.

The term **construction project** is defined as a set of the seven essential actions needed to produce a new facility which have a start and end date. They are producing a brief, design and plan, and procurement, manufacturing, production and commissioning.

Most practitioners think about their work exclusively in terms of projects. This is a practical response to construction being undertaken on distinct sites which have individual requirements and often pose unusual problems. Ground conditions, local environments and site access often face practitioners with issues which fall outside their own direct experience. They therefore concentrate on the immediate project. However, the resulting project-based culture gives rise to several major weaknesses. Those responsible for construction projects tend to be faced with new situations and so are continually applying basic knowledge and skills. Even where professionals and craftsmen have learnt to work together efficiently on one project they may never work together again as they all move to different new projects. Ideas developed on one project are often lost when those involved begin other projects which pose different issues and challenges. These and many other similar weaknesses make it difficult to justify long term investment in the construction industry. As a result many construction companies concentrate on being flexible enough to tackle a wide range of projects. Efficiency is sacrificed in the interests of short-term profits.

One way of overcoming some of the weaknesses of a narrow focus on individual projects is to organize construction programmes. A **construction programme** is a series of construction projects which have sufficient characteristics in common to justify treating them as an integrated whole. Many experienced customers organize their demands on the construction industry into construction programmes. Construction companies that undertake complete developments to produce new facilities which they sell or lease to other organizations, arrange their work in construction programmes. The most obvious and immediate benefit is that construction teams involved together on project after project develop efficient ways of working. It makes economic sense to invest long term in innovation and training. Major investments in plant and equipment make financial sense. Formal approvals and legal processes can be streamlined. Construction management is at the heart of identifying, developing and exploiting these opportunities to improve efficiency.

The benefits of organizing construction in ways which justify long-term thinking and investment lead some groups of companies to establish formal legal relationships. These range from fairly loose agreements to work together whenever an opportunity arises to undertake a construction programme through to formal agreements to merge individual companies into a new company able to provide a complete construction service to customers. As a result, construction in practice has adopted a number of distinct approaches. The most significant are described in Chapters 5 to 9.

3.6 Construction organizations

Construction actions are undertaken by a great variety of organizations. These are based on professions, crafts, specialist actions and various financial, legal and commercial characteristics. The types of organizations involved have changed over time and tend to vary from country to country. They are also different for distinct types of constructed facilities reflecting the technologies involved in, for example, housing, commercial building, transport infrastructure, processing plants, and so forth.

The great variety of organizations undertaking construction actions makes it impractical to deal with them individually in theoretical terms. This means the theory of construction management requires the complexity of practice to be simplified by identifying and defining categories of construction organization which are important in understanding construction management.

Most construction organizations involve more than one individual and some comprise many hundreds or thousands of individuals. However, the term construction organization also includes the special case of one individual undertaking a construction action independently.

The first distinct type of construction organization is the **construction team**. This is a formal group of individuals who work together on a permanent basis to undertake specialist construction actions and the essential machines and equipment the team uses. The diversity of specialist construction actions means the machines and equipment needed to support construction teams differs widely from hand tools to many kinds of information technology, excavating machines, tower cranes, concrete pumps and many other powerful and sophisticated machines.

Construction is undertaken by teams because construction actions tend to require the coordinated efforts of several individuals contributing distinct but closely related knowledge and skills. Teams are an important human creation. There is extensive literature on the key roles within teams and the behaviours needed to form and maintain effective teams. Teams need time to develop the most effective ways of working together. For this reason, previous experience of working together is important in selecting construction teams.

The theory of construction management recognises construction teams as the basic organizational unit which undertakes all construction actions. Construction teams form part of construction companies.

A **construction company** is defined as a permanent organization which undertakes construction actions and so supports one or more construction teams. Construction companies provide support for their construction teams in various ways including providing employment, financial security, legal advice, plant and equipment, technical advice, training, management, research and development, and so forth.

Construction companies are as diverse as the construction teams they support. The definition includes them all. This means an architectural practice or an engineering design firm is defined as a construction company just as much as a general contractor, steel erection contractor, roofing contractor, heating engineering contractor, electrical contractor or decorating firm.

Construction companies vary in size from very small, local companies specialising in a narrow range of actions to huge companies operating internationally and usually undertaking a great variety of construction actions.

Construction companies, apart from the very smallest, are organized into divisions. A **construction company division** is a distinct organizational part of a construction company. The basis for individual divisions varies and includes types of constructed facility, geographical regions, particular construction actions and specialist support for construction teams. Consequently companies may have divisions responsible for superstores, housing, construction in Scotland, cost planning, finance, legal issues, and so forth.

Construction companies form the **construction industry**. It is defined as the totality of construction companies currently operating. It is often useful to refer to some division of the construction industry. These sub-categories may be geographically based, for example, the London construction industry, the United Kingdom construction industry, Europe's construction industry, and so forth. There are other bases for identifying sub-categories of the construction industry including the type of facilities produced, for example, the social housing construction industry, the commercial office construction industry, the motorway construction industry, and so forth. In each case the term includes all the construction companies which meet the sub-category criteria.

Construction teams are not only part of construction companies, they also form part of construction project and programme organizations. A **construction project organization** is the set of construction teams responsible for undertaking a construction project. Taken together they need to be capable of undertaking the complete set of seven essential construction actions. A **construction programme organization** is the set of construction teams responsible for undertaking a construction programme. They form construction project organizations to undertake each project. They have a strategic responsibility for ensuring that linking projects delivers benefits in terms of greater efficiency than would be achieved by tackling the projects individually.

This completes the description of **construction organizations** which are defined as formal groups of individuals who undertake construction actions. Thus, the term includes construction teams, companies, company divisions, project organizations and programme organizations. As Figure 3.1 suggests construction companies, their divisions and construction teams form systems which are like neural networks as they combine to undertake construction projects and programmes.

Construction is influenced by organizations outside the construction industry. As described earlier, the role of customers is crucially important to construction but they are not part of the construction industry. However, some construction is initiated by construction companies that make a business of developing new facilities and then selling or leasing them. In terms of the theory of construction management, development companies act as customers while they are initiating construction actions; when they are undertaking construction actions, they are a construction organization. Development companies have customers who buy or lease their facilities but in terms of the theory described in this book, they are not construction customers.

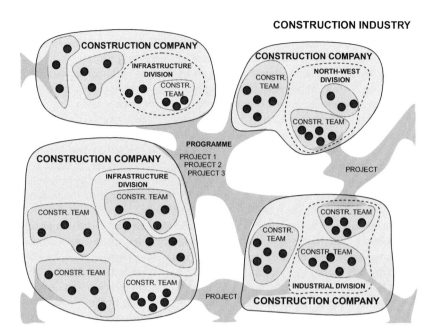

Figure 3.1 Construction Organisations.

In addition to customers, other organizations outside the construction industry influence construction. This is because construction is a significant sector of almost every national economy, employing millions of people, requiring massive capital investments and providing an important component of gross national product. It is therefore inevitable that various external organizations seek to influence construction. These include all levels of government, local communities or organizations affected by new proposals for construction, protest groups, special interest groups and many more. Equally there are national and international bodies which represent construction in its dealings with powerful external organizations. These are not construction organizations as defined in this section but inevitably they influence the practice of construction management.

The ability of external organizations to interfere with construction is taken into account in the theory of construction management without needing to identify the specific organization responsible for the external influence. This is sufficient for the purposes of theory. It is of course essential for practitioners to identify specific external influences. The role of theory is to provide guidance for practitioners on appropriate responses to external interference whatever its source.

3.7 Common characteristics of construction organizations

The theory of construction management concentrates on the characteristics construction organizations have in common. The first important common characteristic is that construction organizations have **information** which is defined as remembered and recorded useful knowledge. A crucial feature of useful knowledge is that it conveys the meanings to everyone using it which are sufficiently similar to provide a basis for effective communication. Ideally everyone would have identical meanings but this unlikely because of the individual influence of prior knowledge.

Knowledge is patterns of relationships between concepts established in our brain. We learn the patterns by categorising our experience and finding consistent relationships. The ability to create new concepts and relationships which are useful improves with experience. This developed ability is called understanding and so useful knowledge is created through understanding.

It is frequently necessary to invent words or limit the meaning of existing words to capture some new understanding. In this way specialised languages develop and where they are found to be useful to a group of people, they become established as a specialised language. This chapter contributes to the development of a specialised language for construction managers which is used throughout the rest of the book to provide useful construction management knowledge.

Knowledge enables us to make deductions about a whole pattern when we encounter a part of it. Thus, if we see a brick wall, a door, several windows and a slate roof, we deduce we are looking at a house and can make logical assumptions about the parts of the house we cannot observe directly. We can go much further and call up a bewildering amount of knowledge because our brain has established multiple patterns of relationships which include the concept of house.

Knowledge found to be useful is remembered and in many cases recorded in some form allowing it to be accessed by others without the originator being present. These records are in books, on the internet, in data bases or some other information technology. The sum total of this useful knowledge constitutes information.

In the context of this book, the information of direct interest is that which enables construction organizations to be efficient and profitable, and survive in competitive markets. Achieving this kind of success requires construction organizations to manage diverse pools of knowledge and understanding. The various ways of undertaking this complex and difficult task are described in Chapters 5 to 9.

The second important common characteristic of organizations is they have **values**. These, most simply, are the positions allocated to things in an order of preferences. All individuals have values. Given a number of options, they rank them in terms of their own preferences. The ranking may well change from time to time and will certainly vary from individual to individual. But at any one point in time, individuals have values and can usually choose between options on the basis of these personal values.

The same is true for organizations but it may be difficult to discover a specific organization's values. Many organizations devote considerable time and resources to establishing their values in respect of, for example, customer service. Carefully designed statements of values are widely published and supported by public statements from senior managers. In practice these are often compromised by internal policies which may inhibit members of the organization from acting on the basis of the stated values. The stated values are frequently contradicted by an organization's actions, especially when faced with problems or crises. Governments are frequently criticised for failing to live up to their own publicly stated values as they react to unexpected events.

Despite the practical failures and contradictions, organizations have values which can be discovered and deduced from statements, policies and actions. The most effective organizations have values which can be seen to run consistently through every aspect of the way they work and interact with others. This consistency is difficult to achieve but is a characteristic which the most successful organizations strive to achieve. The way construction organizations aim to do this is described in subsequent chapters. For the purposes of this chapter it is sufficient to recognise that organizations have values which affect their actions.

The third common characteristic of organizations is they take actions. In the context of this book, we are interested in actions which directly or indirectly produce construction. It is also important to recognise that organizations take actions which are intended and authorised but on occasions they act in ways which are neither intended nor authorised. Important examples include constructed facilities being delivered late, not working properly, including defects and costing far more than anyone ever agreed. An explicit purpose of the theory of construction management is to help practitioners avoid these failures.

3.8 Interactions and relationships

The theory of construction management deals with the effects of relationships between organizations. It does this in ways that help us understand the consequences of common patterns of relationships found in practice. A **relationship** is a linked series of interactions.

The idea that organizations have information and values and can take actions provides the basis for understanding that organizations interact with each other in three distinct ways.

The first important type of interaction is **communication** which is defined as a transfer of information between organizations. Effective communication is important to most human actions and is difficult to achieve due to differences in understanding. This is reflected in the multitude of popular titles in book shops promising to help readers improve their communication skills. The theory of construction management does not directly contribute to this torrent of good, bad and indifferent advice. Its principle contribution to achieving effective communication is to give all those involved in construction a common basis for understanding the factors which determine construction's performance. Also the chapters describing distinct construction scenarios identify the types of information which need to be communicated.

Effective communication requires five features to exist. First, the communicated matter must exist in one organization. Second, the communicated matter is turned into a medium capable of being transferred to another organization. This may be spoken words, marks on paper, electronic data or some other medium. Third, the medium is moved from the originator to the receiver. Fourth, the receiving organization detects the information in the medium. Fifth, the receiver attempts to understand the information. Ideally the originator and receiver achieve a common understanding.

In practice it is difficult to achieve a common understanding because each of the five essential steps in communication can give rise to problems which cause communications to fail and misunderstandings to develop. A common problem in practice is even when the first four stages are carried out competently, the final stage fails because the originator and receiver understand the communicated information in different ways. This is often due to the information being modified by knowledge and understandings already possessed by the recipient; in other words by prejudice.

The important effect is that it is necessary to constantly check the effectiveness of communications and the degree of common understanding. This is important for construction management because many construction problems are caused by failures of communication and the resulting differences in understanding between organizations. When problems are recognised, further communication is needed until all sides accept they have a common understanding of the communicated information.

The second important type of interaction between organizations is even more prone to problems. It is transaction based on values. A **transaction** is an exchange of things of value between organizations. An organization may supply bricks to a second organization in return for money. This transaction will take place if the first organization values the money more than the bricks and the second values the bricks more than the money. If both organizations value bricks and money exactly the same, there is no basis for a transaction because no one can gain. Many of the actions involved in successful transactions take place prior to the actual exchange. The subject of negotiation is devoted to understanding how organizations reach agreement on transactions and providing advice on how to achieve good deals. As with communication, the practical importance of negotiation is amply demonstrated by the multitude of published guides to agreeing transactions.

Construction has elaborate procedures to control the way transactions are agreed. These include professional fee scales, standard forms of construction contract, procedures to guide the production of the information provided to competitive bidders, rules and in many countries laws about competitive bidding, legal minimum wage rates, laws forbidding suppliers to collude in fixing prices, and many other local, national and international conventions and laws. The application of these procedures is influenced by practical considerations which determine whether it is efficient to rely on market forces, restrict negotiations to organizations with an established track record of working together, or keep the transaction inside a single company. The theoretical and practical implications of these key choices are described in Chapters 5 to 9.

Agreeing a transaction involves communication which could be seen as an exchange of information. The distinction between communication and transactions is that communication is concerned with information while transactions involve exchange on the basis of values. Also communication takes place when one organization transfers information to another without any expectation of receiving anything in return. Transactions involve, by definition, an exchange.

In construction, transactions are often a necessary precursor to the third and final important type of interaction. This is the organization of coordinated actions. In effect a new organization is formed because two separate organizations agree to coordinated actions. The new combined organization exists as an identifiable entity while the coordinated actions are being undertaken. It ceases to exist when they are complete. Throughout this process, the two separate organizations remain as distinct entities. There are cases where a transaction results in two organizations merging but this is the unusual in undertaking individual construction actions.

The fact that agreeing to undertake coordinated actions creates a new, usually short-lived, organization means organizations and the distinct parts of organizations interact by means of communication and transaction. This means **interaction** is defined as communication or transaction between organizations.

3.9 Double-loop learning in construction networks

A set of interacting construction organizations and the relationships between them form a **network**. Construction networks include the organizations which undertake construction projects and programmes, and those which run construction companies and their divisions.

These construction networks can all provide examples of a concept which provides important advances in our understanding of the complex organizations. The concept is self-organizing networks which have served to highlight the importance of feedback in enabling organizations to respond and adjust to changes in ways which ensure their survival.

Feedback is information about the effects of an organization's actions. When feedback shows performance is falling outside acceptable limits, organizations which behave rationally modify their actions. This may mean changing their own behaviour, developing new relationships, strengthening existing ones or placing less reliance on those which appear to be the cause of the failure. Then feedback resulting from the modified actions is used to discover the effect of the changes. This may identify the need for further change and so on until feedback shows performance is acceptable. In this way feedback is absolutely vital in enabling complex organizations to learn from their experience and improve their own performance.

The most highly developed organizations establish sophisticated systems which use feedback to drive performance improvements in what is called double-loop learning. This comprises two feedback loops.

- The first feedback loop guides organizations in achieving immediate objectives. In doing so the feedback helps achieve immediate objectives but also may identify actions which will raise the organization's performance levels within its currently accepted norms and standards.
- The second feedback loop operates on a longer time scale to produce ideas which challenge the accepted norms and standards. This feedback identifies actions which improve the accepted norms and standards and enables the organization to raise its performance to an entirely new level.

The concept of double-loop learning recognises that construction teams work together in project organizations under established norms and standards. These influence many aspects of work including the relationships between the teams. The ability to go beyond the first feedback loop, which guides day-to-day work, and achieve double-loop learning depends on the nature of the larger organizations of which the teams are part. Thus double-loop learning depends on support from organizations' strategies and methodologies, construction management approaches, project management tools and techniques, and all the norms and standards which shape the long-term behaviour of teams and their organizations.

Double-loop learning in construction thus provides a framework for the continuous improvement of performance and the development of new and innovative approaches to construction management. Chapters 5 to 9 describe the major construction management approaches and show how some inhibit while others support double-loop learning. These later chapters show the most important condition for double-loop learning to be achieved is the consistent use and review of systems and procedures over time. This can only be achieved by construction teams that repeatedly work together on project after project. The crucial importance of permanence permeates this book and forms a fundamental building block of efficient construction management.

3.10 Categories of relationship

The performance of construction project organizations is influenced directly by the quality of relationships between the construction teams involved. In practice the relationships between construction teams vary greatly and so form a multi-facetted range of relationships.

The best way to describe the various categories of relationship is to consider any kind of construction organization as a system comprising components linked by internal and boundary relationships as illustrated in Figure 3.2.

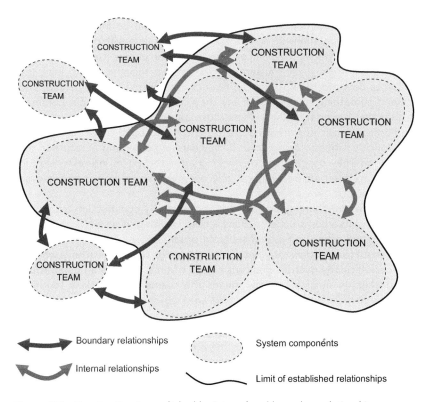

Figure 3.2   Construction teams linked by internal and boundary relationships.

At one extreme are **boundary relationships**: relationships in which construction teams' behaviour is guided by their perception they are parts of different organizations. Teams working together for the first time and employed by different companies are likely to have formal boundary relationships. Typically these relationships are based on terms and conditions agreed in an individual transaction.

Relationships based on individual transactions are influenced by the nature of the negotiations needed to agree the transaction. This can be a lengthy series of interactions in which the parties build confidence in each other, develop a common basis for communication, and agree the precise terms and conditions of the transaction. If both parties regard the negotiations and the resulting terms and conditions as fair, relationships between their construction teams are likely to be effective and transform over time into established relationships. However, if either party regards the negotiations or the terms and conditions as unfavourable to their interests, subsequent relationships between their construction teams are likely to be acrimonious and inefficient.

Once the transaction is agreed, companies and their construction teams have to develop sufficient common ground for them to communicate and work together effectively. This can involve the exchange of formal documents, much correspondence, formal meetings, informal meetings, casual meetings and social events. Misunderstandings will arise, problems in the original agreement will be identified, and progress towards effective communication and action is likely to be slow. The parties have to rely on local conventions, language and terminology which may or may not be appropriate for their particular relationship.

Once the construction teams begin work, they will be influenced by the knowledge their relationship is for a fixed length of time with pre-determined start and finish dates. This encourages them to ensure they individually earn a profit from each transaction. To achieve this they may: limit the resources devoted to joint actions; restrict the information they share; and demand something in return for any they do provide. The most common result is that boundary relationships cause parties to stick to the terms and conditions set out in formal contracts. They help each other only if they are explicitly required to do so or if they expect something in return in the short term. They concentrate on looking after their own interests, exploit problems for their own advantage, and do everything they can to minimise their costs.

The process of forming boundary relationships by construction teams undertaking related construction actions is called **folding**. It results in the teams remaining and acting as parts of separate organizations.

In contrast to boundary relationships, construction teams which have worked together efficiently over a significant period of time develop an **established relationship**. In most cases this means the teams have worked together efficiently on a number of projects. Where an established relationship continues over many projects and a number of years, it is likely to develop into a high quality **internal relationship**. This is a relationship in which construction teams' behaviour is based on the perception they are parts of a joint organization. They share information and resources whenever this is expected to benefit the joint organization. They share a well-developed, common language and effective communication systems. They have established ways of allocation work widely regarded as fair. They use well-developed technologies in which the construction teams are skilled. They help each other by spotting problems and cooperating in solving them.

High quality internal relationships are most likely when construction teams are part of the same division of a company. When teams which have worked together efficiently are parts of different divisions within a company, provided relationships between the divisions are harmonious, good quality internal relationships are likely. Even when teams which have worked together efficiently on a number of projects are parts of different companies, their previous experience is likely to enable them to form reasonably good quality internal relationships. In all these situations the teams begin related actions quickly and are able to work efficiently. They concentrate on their work safe in the knowledge they will not be delayed or interrupted by the others and they achieve high levels of efficiency.

The process of developing internal relationships is called **composing**. Construction teams involved in composing are continuously changing and improving their relationships. The definition is not influenced by the formal employment situations of the teams or the terms of the contracts which bring them together.

The differences between folding and composing are particularly evident when teams move from one project to another. In folding, teams always behave in the same way and ultimately resort to contractual obligations and make no attempts to improve relationships with other teams. In composing, on the other hand, construction teams have established relationships from previous projects and aim to continuously improve the quality of their relationships.

The **quality of relationship** is a measure inversely related to the time and resources devoted to interactions by construction teams. High-quality established relationships enable construction teams to devote the great majority of their time and resources to their construction actions and a minimum of time and resources to interactions. In contrast construction teams working with low-quality boundary relationships devote far more time and resources to interactions; consequently they are less efficient at undertaking their construction actions. Figure 3.3 shows a range of relationships and the likely outcomes in terms of efficiency.

3.11 Factors influencing construction performance

There are two factors, in addition to those already described in this chapter, which influence construction performance and may cause greater uncertainty. The first is **performance variability** which is a measure of the range of performance achieved by construction teams.

It is part of the human condition that we do not always achieve exactly what we set out to do. Some days we race through our work but on others we seem if anything to be going backwards with more to do at the end of the day than we knew about at the beginning. Construction teams have these normal human characteristics and their performance varies from day to day and week to week. This needs to be taken into account by those responsible for construction projects. They may plan for a particular action to be completed on a given day but it may be finished early, on time or late.

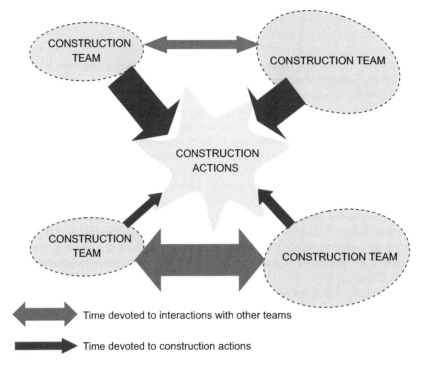

Time devoted to interactions with other teams

Time devoted to construction actions

Figure 3.3 Quality of relationships.

Many factors influence the level of performance variability experienced on any particular construction project. These include the competence of the construction teams, their experience of similar work, their experience of working with the other construction teams involved, the effectiveness of communication systems, and so forth. Research has established measures of performance variability for many construction actions. Even on well-run projects it can be of the order of ± 50 to ± 60 per cent of established industry norms. The range can be much wider on projects experiencing problems.

Performance variability influences construction performance by making it difficult to plan with reasonable certainty. Resources are wasted when one team is delayed by a failure on the part of another team. Construction teams may have expensive plant and equipment stand idle or work in a sub-optimal manner. It is therefore a factor which the theory of construction management deals with directly.

This is also the case with the second factor which influences construction performance. It is **external interference** which is a measure of the impact of factors external to a construction project.

Construction is a complex process and this makes it vulnerable to many external factors. In the early stages of construction projects these tend to emanate from powerful organizations. These include government bureaucracies, politicians, special interest groups, competitors and increasingly various forms of the media. As construction projects move towards the manufacturing and production actions, different factors may compromise performance. These include all the problems which affect manufacturing and transport systems, difficult weather conditions, industrial disputes, market conditions, labour supply and many other factors external to the project organization.

The distinguishing characteristic of external interference is it is caused by factors which affect construction performance but are outside the direct control of the construction project organization. The level of external interference experienced by any construction project is difficult to predict and in practice because of this uncertainty, construction project organizations devote considerable effort to risk management with the aim of insulating their projects from external factors.

3.12 Construction management

Having defined the key concepts involved in understanding construction performance, it remains to consider construction management. The general definition of **management** is it means taking responsibility for the performance of an organization.

Everyone who forms part of an organization is responsible for ensuring its performance is efficient. This means management is undertaken by teams responsible for directly producing the organization's outputs and by specialist managers. Managers exist because it has been decided that specific actions required for an organization to work efficiently should be undertaken, not by teams directly producing the organization's outputs, but by specialist in those actions. Managers, like everyone else, are responsible for their own actions but unlike most specialists they are also responsible for the actions of others. This is the distinguishing characteristic of management.

It follows that **construction management** is defined as taking responsibility for the performance of a construction organization. Fairly obviously performance depends on the competence of the construction teams, the quality of their relationships, and the levels of performance variability and external interference they experience. However, their performance is also dependent on the competence of the construction project organizations, programme organizations, companies and company divisions of which they form parts, and by the quality of the relationships between these organizations.

Construction managers exist because it has been decided that specific parts of the essential construction actions should be delegated to specialist managers. This does not in any way modify the vital principle that construction performance is the responsibility of everyone involved. Managers, like everyone else, are responsible for the actions delegated to them; an important part of this is being responsible for the way their actions influence the performance of all the other teams involved.

3.13 Construction efficiency

The theory of construction management is intended to guide practitioners in making decisions which help them achieve their objectives. It follows that for the purposes of the theory of construction management, **efficiency** is a measure inversely related to the waste caused by complexity and external interference which prevent organizations achieving their agreed objectives.

It is important for practitioners and their customers to have more objective measures of the performance of construction. Such measures are also vital for researchers to identify the strengths and weakness of alternative construction management strategies. The practical issues involved in measuring construction performance, including efficiency, are discussed in Chapter 11.

The concept of efficiency as defined above provides a sufficient basis for the theory of construction management which is described in the next chapter.

Exercise

The following tasks are designed to help readers understand and use the concepts described in this chapter.

1. Suggest the types of construction organizations likely to be involved in each of the seven essential construction actions.
2. Identify some specific construction organizations, list their main characteristics and identify characteristics they have in common and those which are unique among the selection of construction organizations.
3. Describe the main features of boundary relationships and internal relationships likely to be found between each of the following pairs of organizations.
   - Customer and design consultant.
   - Architect and services engineer.
   - Design team and management team.
   - General contractor and bricklaying team.
   - Electrical installation team and ceiling installation team.

Further Reading

The following publications are the source of ideas used in this chapter and provide further information for readers.

Axelrod, R. (1984) *The Evolution of Cooperation*. Basic Books. This book provides a crucial step in the development of ideas about cooperation. It describes many natural and human situations in which cooperation has emerged and identifies the principles which explain why this happens. It shows that human organizations are most successful when people realistically expect each other to act cooperatively.

Capra, F. (1996) *The Web of Life*. Harper Collins. This book builds on developments in scientific descriptions of evolution to describe how cooperation and symbiosis have been central to the evolution of life on Earth. It sees the world, including all living creatures, as one incredibly complex system of networks in which feedback loops give the whole and individual parts the power of self-organization. The book also identifies the key characteristics of networks and importantly explains how complex systems can be viewed at distinct levels which have properties which do not exist at other levels.

Morgan, G. (2006) *Images of Organization*. Sage. This powerful and insight-
ful book details what exactly is happening in organizations. The author's
predominant aim is to reveal the power of metaphor in shaping organization and is therefore based around the implications of different metaphors on the organizational life. The metaphors include organizations as
machines, organisms, brains, cultures, political systems, psychic prisons,
flux and transformation, and instruments of domination.

Chapter Four
Theory of Construction Management

4.1 Introduction

The theory of construction management begins with the proposition that construction management aims to enable construction to be undertaken efficiently.

The first and most obvious requirement in achieving this objective is to select competent construction teams. This means the teams understand and are experienced in the construction actions they are required to undertake. The most straightforward way to ensure this is for construction management to select teams they have worked with before in successfully undertaking similar construction actions in similar circumstances. Where this is not possible construction management should identify a number of potentially competent teams and seek robust evidence of each team's track record. A crucial part of any team's track record is the support provided by the company which employs them. There are various sources of such information including customers, construction managers and other construction teams the team has worked with on previous projects. The most useful information is based on recent projects. This can be supplemented by information about qualifications, training and experience. The selection process can include competitive submissions and interviews. In cases where the required construction is critical, it may be sensible to devise tests of a team's competence before making a final selection. The importance of selection processes is reflected in the extensive construction management literature dealing with the selection of competent construction companies and teams. The following description of the theory of construction management assumes this well-developed guidance is followed and competent teams are selected to undertake all the essential construction actions.

Given competent construction teams, construction management is centrally concerned with ensuring the teams, which need to interact, establish effective relationships. The importance of interactions is illustrated by the very simple example described in the box.

Construction Management Strategies: A Theory of Construction Management, First Edition. Milan Radosavljevic and John Bennett.
© 2012 John Wiley & Sons, Ltd. Published 2012 by John Wiley & Sons, Ltd.

The Hénon attractor

Complex behaviour can arise from very simple deterministic systems even though it is often mistakenly thought it results only from extreme levels of complication. The Hénon attractor is a good example of complex behaviour arising from a dynamical system of just two variables.

$$x_{n+1} = y_n + 1 - ax_n^2$$
$$y_{n+1} = bx_n$$

The behaviour of the above system depends on just two parameters a and b. The system behaves chaotically when $a = 1.4$ and $b = 0.3$ but may also behave chaotically for other values. Plotting x_{n+1} against y_{n+1} for the above values of a and b reveals an infamous horse-shoe diagram depicting the chaotic Hénon attractor. Interestingly not a single classical statistical test used on x_{n+1} and y_{n+1} would detect a deterministic underlying system. The Hénon map is a time-dependent system where values of x and y in step $n + 1$ depend on the values obtained in step n. This is also the main reason we have used the Hénon attractor as an example. Figure 4.1 shows in very clear terms that subject to appropriate conditions even such a simple time-dependent dynamical system can exhibit hugely complex behaviour.

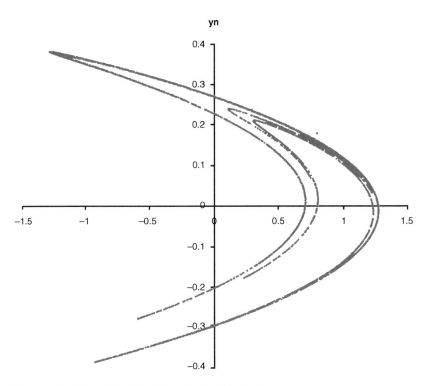

Figure 4.1 Hénon Attractor with $a = 1.4$ and $b = 0.3$.

> Construction is far more complex. Instead of dealing with just two variables construction projects have a number of interacting teams where outcomes in the future depend on the number of involved teams, the quality of relationships between interacting teams and their performance variability. In addition there is also unpredictable interference which may arise from numerous external factors which form an additional set of parameters and make construction inherently difficult. Such a level of complexity and external interference requires highly effective construction management.

As explained in the box, construction management normally faces systems which involve significantly more interactions than the Hénon attractor. This is illustrated by considering a construction project which gives rise to an extreme level of inherent difficulty. It is a worst case scenario. Such construction projects are rare but they can occur.

4.2 A worst case construction project

A construction project faced with extreme inherent difficulty involves many organizations linked by complex, contradictory, unclear and disputed contracts to undertake the following actions.

- Set up complex organizational arrangements to produce a **brief**. The result is a complicated, incomplete and contradictory description of highly sophisticated additions and alterations to the built environment. The complex organization frequently changes its composition and methods and continually revises and alters the brief in unpredictable and arbitrary ways.
- Produce, revise and repeatedly alter complex and incomplete **designs** for the product of construction actions which are beyond the competence of available construction companies. The designs incorporate many sophisticated innovations and new technologies, require extensive new research and development and only partially satisfy the requirements of the current brief.
- Produce complex, contradictory and incomplete **plans** which at best only partially satisfy the requirements of the current brief and complicate and confuse all the design, procurement, manufacturing, production and commissioning actions by, among other problems, requiring actions outside the competence of available construction teams.
- **Procure** construction teams which lack the required skills, equipment and facilities and lack the motivation to undertake the construction actions described by the design within the framework described in the plan. The procurement actions are undertaken late and result in construction teams being forced into accepting tough, unenforceable contracts which give rise to delays and disputes. Materials and components, only some of which are necessary, are procured late, wrongly delivered to site, mislaid and lost.
- **Manufacture** the individually designed components and materials which require new research and development, complex design and innovative manufacturing process which are beyond the established competence of available manufacturing companies.
- Undertake **production** using inappropriate skills, equipment and facilities in ways which waste and damage materials and components. The outputs

constantly need to be redone and altered and differ markedly from the current versions of the designs and plans.

■ **Commission** the alterations and additions to the built environment in a piecemeal manner which delivers an incomplete facility which never functions properly and is very different from that described in the current brief leaving the organization which takes over and runs the new facility incapable of using or operating it effectively.

All the actions rely solely on contractually determined boundary relationships within an ill-defined and constantly changing project organization. The construction companies which supply the construction teams rely solely on vague and imprecise contractual terms and conditions. As a result, they engage in disputes, resort to legal sanctions and act in wholly competitive and entirely uncooperative ways. The whole project organization is beset by massive and repeated interactions with the organizations' hostile, rapidly changing and unpredictable physical and organizational environments.

The worst case scenario is now illustrated and expressed in mathematical terms to provide a more precise description of the construction management task. Figure 4.2 illustrates construction companies comprising divisions, teams and internal relationships which contribute to a construction project organization and in doing so form a complex network comprising teams and relationships. In the worst case scenario the project organization is formed through folding (**F**) which is a term borrowed from general systems theory. It refers to the formation of a higher order system from a collection of subsystems without changing the quality of relationships that exist in the subsystems. In relation to a project this would correspond to forming a project organization where individual teams limit their relationships to contractual obligations and make no attempt to achieve positive synergy with other teams.

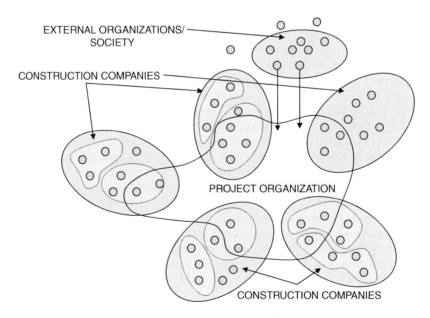

Figure 4.2 A Worst-Case Construction Project Network.

This scenario results from a complete absence of any previously established relationships between the construction teams. To represent this more clearly, Figure 4.2 shows solid boundaries around different construction teams to indicate their unwillingness to collaborate and over-reliance on contracts. In the absence of established relationships these fundamental boundary elements strengthen the boundaries even further. Dotted lines on the other hand represent boundaries between teams with established relationships where the boundary only identifies the existence of separate teams. In this scenario, such established relationships exist only inside individual construction companies and provide little or no benefit to the project organization.

The above construction project network can be expressed mathematically using the general system theory. The fundamental assumption we need to make is that the network is a complex system of n-th order because it is created by teams from participating companies, which are systems of higher order. Such a system is a collection of subsystems which can be expressed as:

$$S^{(n)} = S^{(n)}\left(C^{(n)}, R^{(n)}, BR^{(n)}\right) \qquad [4.1]$$

$C^{(n)}$ is a set of components of n-th order (e.g. individual teams that form a system of n-th order).

$R^{(n)}$ is a set of relationships of n-th order (i.e. relationships that exist between individual teams).

$BR^{(n)}$ is a set of boundary relationships of n-th order (e.g. relationships between individual teams and the external environment).

To form the above system, we need a collection of systems of lower order. For simpler annotation we assume that all these systems are of $(n-1)$-th order but it should be understood that they can be of any order below the newly created system. Let therefore $S_i^{(n-1)}$ be the i-th team in the project organization which comes from one of the participating permanent organizations:

$$S_i^{(n-1)} = S_i^{(n-1)}\left(C_i^{(n-1)}, R_i^{(n-1)}, BR_i^{(n-1)}\right); \quad i = 1, 2, \ldots, m \qquad [4.2]$$

We then fold these teams in an open system of the n-th order:

$$S^{(n)} = \overset{m}{\underset{i=1}{F}} S_i^{(n-1)} \qquad [4.3]$$

$$S^{(n)} = S^{(n)}\left(C^{(n)}, R^{(n)}, BR^{(n)}\right) \qquad [4.4]$$

The set of components equals the union of sets of components of the first order systems (i.e. a collection of all teams from participating permanent organizations):

$$C^{(n)} = \bigcup_{i=1}^{m} C_i^{(n-1)} \qquad [4.5]$$

The set of its internal relationships includes the sum of internal relationships of the original teams and their new relationships (i.e. relationships between participating teams):

$$R^{(n)} = \left(\bigcup_{i=1}^{m} R_i^{(n-1)}\right) \cup SR^{(n)} \qquad [4.6]$$

$SR^{(n)}$ represents relationships between participating teams.
$S^{(n)}$ is a system of order 2 (constituted by its components, internal relationships and boundary relationships)

A system of n-th order thus comprises $(n-1)$-th order systems which interact with each other through what can be best described as internal boundary relationships. Intrinsically, the system of higher order retains the relationships that were already present in the original systems. In other words, the quality of relationships after folding systems of first order into a system of second order does not change. In practical terms this corresponds to a failure of teams which come from permanent organizations to form any real allegiance to the project seeking only their individual advantage, and sticking narrowly to their contractual obligations.

Within the worst case scenario, the project organization is a temporary system of higher order which is formed by the participating permanent organizations for a limited period of time. It brings together teams which have never worked together before and they cannot be expected to form effective relationships within a short period of time. This is why they do not change and the components of the n-th order system are simply a collection of unchanged teams from participating companies. There is no willingness to pursue the synergistic effects of collaboration and they base their involvement solely on contractual obligations. Although such behaviour is counterproductive, it is a direct result of the lack of established relationships among a large number of involved construction teams and the absence of incentives to develop effective relationships. Folding is the inevitable outcome of construction projects which are close to the worst case scenario.

4.3 A straightforward and certain construction project

To emphasise the nature of the construction management task a second case is described. This is an ideal case provided by the most straightforward and certain construction project. It involves one organization which organizes directly employed and fully competent construction teams who are well experienced at working together on similar projects to undertake the following actions.

- Produce a **brief** which provides a clear, straight-forward and complete description of a new facility and do not subsequently change the brief.
- Produce a clear, complete **design** for the product of construction actions which satisfy the requirements of the brief and do not subsequently change the design.
- Produce a clear, complete **plan** which ensures design, procurement, manufacturing, production and commissioning actions can be undertaken efficiently and safely, at times, costs and quality standards which satisfy the requirements of the brief and do not subsequently change the plan.
- **Procure** the necessary materials and components which are all readily available from the routine output of manufacturing organizations.
- Undertake **production** using the required skills, equipment and facilities to convert the materials and components into the required addition or

alteration to the built environment precisely in accordance with the design and plan.

■ **Commission** the addition or alteration to the built environment to produce the fully tested and properly functioning new facility described in the brief and ensure the organization which takes over and runs the new facility is trained to use and operate it.

These actions involve highly effective internal relationships within a composed organization and are undertaken without significant interaction with the organization's benign physical and organizational environments.

This is now illustrated and expressed in mathematical terms to provide a precise description of the relationships which support the construction management aim of enabling construction to be undertaken efficiently.

A single organization exists which comprises all necessary construction teams to achieve the agreed objectives. Only a very small proportion of minor specialist works would be outsourced in this scenario posing very limited contractual risks. There is a single contractual relationship between the organization and the customer and very few other contractual relationships as illustrated in Figure 4.3. The vast majority of relationships have been established during previous projects because the teams come from the same organization. This has the important benefit that relationships are not inhibited by contractual limitations.

It is important that highly effective relationships take time to develop and shortcuts in the form of mergers and acquisitions cannot replace organic growth. The benefits of this scenario are not simply based on teams being part of the same legal entity. They depend much more on teams working under the same umbrella organization for many years, overcoming the potential dangers of departmentalisation, developing close ties and building effective relationships.

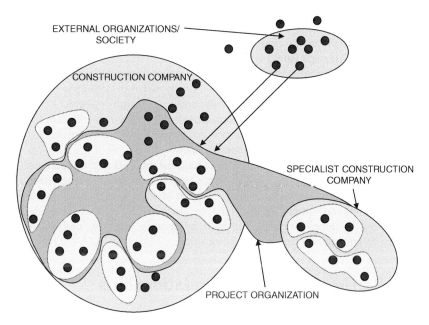

EXTERNAL ORGANIZATIONS/
SOCIETY

CONSTRUCTION COMPANY

SPECIALIST CONSTRUCTION
COMPANY

PROJECT ORGANIZATION

Figure 4.3 A straightforward and certain construction project network

This scenario includes construction teams which have developed effective relationships over many years and through numerous projects so the project organization results from the composition of subsystems into a system of higher order. Composition is another term that has been borrowed from general systems theory and refers to the creation of new relationships. With the composition **C** we make a new system of *n-th* order in the following way:

$$S_*^{(n)} = \overset{m}{\underset{i=1}{C}} \, S_i^{(n-1)} \qquad\qquad [4.7]$$

such that:

$$S_*^{(n)} = S_*^{(n)}\left(C^{(n)}, R_*^{(n)}, BR_*^{(n)}\right) \qquad\qquad [4.8]$$

where

$S_*^{(n)}$ represents a new system of higher order.
$C^{(n)}$ is a set of components of the higher order system (they obviously do not change).
$R_*^{(n)}$ is a changed set of internal relationships in the higher order system.
$BR_*^{(n)}$ is a changed set of boundary relationships in the higher order system.

It is evident that we are talking about the same teams before and after composition but the relationships change. This is in essence a key difference between the two processes (*folding* and *composing*). Composition brings relationships into a system of higher order, such that:

$$R_*^{(2)} \neq R^{(2)} \qquad\qquad [4.9]$$

$$BR_*^{(2)} \neq BR^{(2)} \qquad\qquad [4.10]$$

Composition is a process which generates a new entity with new characteristics and creates conditions which enable the new entity to continue to change the quality of relationships within and between constituent components.

This explains why teams with established relationships continuously evolve and develop the way they work together. Composition provides the synergy which drives a continuous change of relationships through repeated interactions spanning many years and numerous projects. This is very different from situations created by competitive selection processes which often prevent the emergence and maintenance of effective relationships.

Composition is most often found inside long-established construction companies capable of undertaking complete projects. Nevertheless, it is possible to establish similarly effective relationships in cases where a construction company establishes long-term relationships with specialist construction companies. This is achieved by employing the specialist companies regularly and actively seeking to enhance the resulting relationships.

4.4 Barriers to effective relationships

The task of establishing effective relationships in a manner which ensures construction is undertaken efficiently is challenging because construction can create barriers which restrict the behaviour of even the most competent teams. It is therefore essential that construction management understands the nature of the inherent difficulty formed by these barriers in order to devise strategies which remove them or at least reduce their impact on efficiency.

The following descriptions concentrate on the level of individual construction projects because this provides a convenient basis for describing inherent difficulty. It is important for construction management to recognise that all construction organizations are affected by inherent difficulty.

There are three fundamental sources of inherent difficulty in construction projects. The first is the design. The second is the construction teams employed to undertake the project. The third is the environment which influences the project.

4.5 Inherent difficulty caused by design

The design of a facility determines the number of distinct technologies involved in its construction and how closely they are interrelated. This in turn influences the number of separate construction teams required to undertake the required work and determines the need to establish relationships between them.

These two key characteristics directly influence the complexity of the construction project organization which in turn contributes to the inherent difficulty it faces. Construction management strategies exist to mitigate the inherent difficulty actually experienced by a project organization working with any given design. More advanced construction management strategies aim to influence the design in ways which reduce inherent difficulty at source. These are described in Chapters 6 to 9.

For the purposes of this chapter it is necessary to describe how the design of a facility influences the number of construction teams involved in its construction and the interactions between them.

Distinct technologies result from the application of knowledge and skills which provide the practical basis for individual construction teams. In practice there is no fixed relationship between knowledge and skills, and the number of construction teams. For example, a design team may specialise in the design of external cladding. Another design team may specialise in the design of supermarkets, while yet another designs the layout of new residential estates or even new towns. Some design teams are based on aesthetic knowledge and skills while others use engineering, acoustics, thermodynamics, communications or any one of the many other bodies of specialised knowledge and skills required to design modern facilities. Similarly one production team may specialise in installing the formwork needed for vertical reinforced concrete columns, while another may specialise in the formwork for complete structural frames, while yet another produces complete reinforced concrete structural frames. In yet a further variant, other production teams undertake a wide range of reinforced concrete work including foundations, ground slabs, structural frames, staircases, walls and individual columns and beams.

The effect of this practical variety is design determines the knowledge and skills required in undertaking a construction project but this merely establishes the starting point for decisions about the type of construction teams to be employed. As illustrated by the practical examples given above, any given set of technologies can be delivered by many different combinations of construction teams.

Local construction industries establish norms which determine the type of construction teams employed to realise broad categories of design. Thus on one project, established local practice may result in one production team producing the complete reinforced concrete foundations, ground slabs, structural frame, staircases and walls for a construction project. This may well mean the team needs few interactions with other teams and those which are required are simple and straightforward. A different category of design in the same local construction industry may result in a larger number of construction teams and many relationships.

So the design, interpreted in terms of the conventions existing in the relevant local construction industry, provides a starting point for determining the complexity of a construction project organization. This complexity is a major factor in determining inherent difficulty.

4.6 Inherent difficulty caused by construction teams

The construction teams involved in a construction project undertake the required work to the level of efficiency established by the local construction industry. The level of efficiency can be changed over time but in the short term, which means the life of all except large construction projects, it is effectively fixed in any given local construction industry. So this has a very direct influence on the level of efficiency which can be achieved; but it does not directly affect inherent difficulty.

Construction teams' influence on inherent difficulty comes from them undertaking their work at an inconsistent pace. Output varies from day to day even when the surrounding conditions are very similar. It follows that construction actions are characterised by variability which is a characteristic of all seven fundamental construction actions. Variability directly influences the complexity of the construction project organization which in turn determines the inherent difficulty it faces.

As described in Chapter 3, construction teams have good days and bad days. This is measured in terms of variability around the norm. Thus a bricklaying team which on average lays 2,000 bricks a day may in practice lay anywhere from 500 or less to 3,000 or more in a day. This makes planning the time for the production of brick walls problematic. The resulting uncertainty raises problems for other production teams which need the walls to be complete before they can undertake their work. These problems can rapidly escalate into major difficulties on projects which involve many closely interacting construction teams. Work is delayed, costs rise and almost inevitably quality suffers as construction teams have to work in less than ideal situations.

The variability of briefing, designing, planning and procurement actions is often greater than manufacturing, production and commissioning. It is often difficult to predict when a complete and fully agreed brief will be produced. Customers may experience major changes in their own business and have to revise their requirements. An entirely new approach to a proposed facility may suddenly appear irresistibly attractive to a customer even if a brief has already been agreed. A design problem can delay a project for months while alternative technologies are tested, new development work is undertaken and in extreme cases new research is carried out. Initial plans may prove to be impractical because of local circumstances, unusual features included in the design or a revised brief. Initial efforts to procure key members of a construction project organization may fail, bringing all actions to a halt while a complex selection process is repeated. These and many other similarly unfortunate events cause variability which is the second source of the complexity of construction project organizations and therefore of the inherent difficulty they face.

4.7 Inherent difficulty caused by construction environments

The third source of inherent difficulty is the impact of factors external to a construction project which are outside the control of the construction project organization. As Chapter 3 describes, factors of almost every imaginable type may interfere with the progress of a construction project. An engineer carrying the only current version of crucial design calculations in his head may be killed in a train crash. Local residents may bribe a local government official to ensure permission to build is refused. The architect's offices may be subject to a prolonged power cut. A factory manufacturing key components may be shut by an industrial dispute. Key components being delivered may be damaged in a road accident. The steel frame of the facility may be struck by lightning during construction. The construction site may be covered by several metres of snow. An environmental protest group may target the construction site bringing work to a halt. A new construction project may begin work nearby and attract all the available labour by offering better wages and working conditions.

All those experienced in construction can add many other examples of external factors which have impacted projects in unexpected ways. All construction is open to the risk that an external factor will interfere with progress. This external interference is the third source of the inherent difficulty faced by construction projects.

4.8 Inherent difficulty

Inherent difficulty is a measure of the complexity and external interference experienced by a construction project organization using traditional construction practice. That is, locally established practice before any construction management strategy is applied.

Complexity is a measure of the number of interacting construction teams involved in a construction project, the quality of the relationships between them and their performance variability. Complexity increases directly with the number of construction teams involved, inversely with the quality of relationships, and directly with performance variability.

External interference is a measure of the impact of factors external to a construction project. It is normally expressed as the percentage of time construction teams are delayed by external factors. Inherent difficulty is directly related to the occurrence and amount of delay.

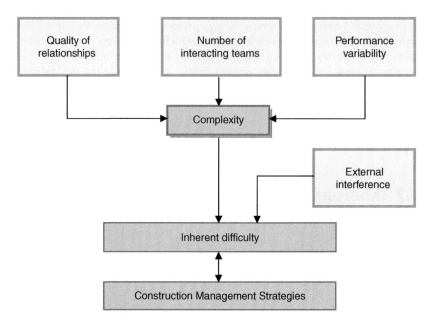

Figure 4.4   Inherent Difficulty of Construction Projects.

Figure 4.4 shows the component parts of inherent difficulty and suggests its direct influence on construction management strategies.

4.9 Construction management

The theory of construction management accepts construction faces inherent difficulty which is hard to avoid. Nevertheless, the theory is based on the rigorous view that the purpose of construction management is to reduce inherent difficulty.

This means in theory there are sets of construction management decisions which enable construction to be straightforward and certain. This view establishes the primary purpose of the theory of construction management, which is to guide practitioners' decisions towards making construction straightforward and certain and so enabling it to be undertaken efficiently.

4.10 Construction management strategies

The situation faced by any individual construction project is shaped by characteristics of the project, the companies which provide the construction teams, and the project's environment. In the absence of any construction management strategy, the project will adopt **traditional construction** practice. This is defined as the locally established actions of construction project organizations. It determines the level of inherent difficulty experienced by any particular construction project in the absence of any construction management strategy and is described in Chapter 5.

Traditional construction practice ensures most construction projects are naturally situated somewhere in the range defined by the worst case, and the straightforward and certain scenarios described earlier in this chapter. The construction management task is to improve on the level of inherent difficulty and guide construction projects as close to the ideal of being completely straightforward and certain as is consistent with the highest achievable efficiency.

The theory of construction management provides a rigorous basis for understanding how this task is undertaken. The theory begins with the propositions that construction projects have a level of inherent difficulty which varies from project to project; and inherent difficulty can be mitigated by construction management strategies.

A **construction management strategy** is a coordinated set of decisions which guide a construction project organization. Construction managers select from a range of strategies to enable them to manage individual construction projects. In general terms the choice is between doing nothing and being more proactive. The proactive strategies involve either directly managing the inherent difficulty or altering the factors which cause the inherent difficulty.

Accepting the inherent difficulty, which means adopting a strategy of doing nothing, is the traditional approach described in Chapter 5. Managing the inherent difficulty can be achieved by using design build, as described in Chapter 6 or management contracting or construction management, as described in Chapter 7. Altering the factors which cause inherent difficulty can be achieved by using partnering, as described in Chapter 8 or an integrated approach, as described in Chapter 9.

In general, construction managers apply the proactive strategies by selecting actions which mitigate the level of inherent difficulty and then matching the actions of everyone involved to the resulting tasks. The choice of an appropriate construction management strategy is shaped by the fact that actions to mitigate inherent difficulty involve costs and benefits. These need to be balanced to determine an optimum set of management actions which means the actions which achieve the highest achievable level of efficiency. This results from balancing the costs of mitigating inherent difficulty with the benefits until a point is reached where the costs of further actions are unlikely to deliver commensurate benefits.

Construction management strategies can be inherent in the work of the construction companies and the teams they supply to form project organizations. Such well-established strategies are highly effective provided they are compatible with the projects undertaken. Other construction management strategies are devised for individual projects while others emerge on an ad hoc basis as projects proceed.

The first step in understanding construction management strategies is to consider the causes of inherent difficulty.

4.11 Basic theorems

The theory of construction management begins with a set of theorems which guide construction management decisions in dealing with the basic causes of inherent difficulty as they aim to achieve the highest possible levels of efficiency. The basic theorems assume construction teams are competent and have the necessary support in undertaking the required actions. There are four basic theorems which are based on the fundamental construction management theorem as shown in 'Basic Theorems' below.

Fundamental Construction Management Theorem
The purpose of construction management is to enable construction to be undertaken efficiently.

Basic Theorems
Efficiency is inversely related to the number of individual construction teams involved in a construction project. Efficiency is directly related to the quality of relationships between the construction teams involved in a construction project. Efficiency is inversely related to the performance variability of the construction teams involved in a construction project. Efficiency is inversely related to the external interference experienced by the construction teams involved in a construction project.

The basic theorems do not directly provide guidance for construction management. They describe the fundamental nature of construction activity from a construction management viewpoint and so guide the propositions about construction management decisions in the next section.

4.12 Basic propositions about construction management decisions

The purpose of construction management decisions is to increase the level of efficiency achieved by construction teams affected by those decisions. To do this, decisions must be consistent with the basic propositions about efficiency. This leads to the following propositions about construction management decisions.

Basic Propositions
Construction management aims to reduce the number of construction teams involved in a construction project. Construction management aims to improve the quality of the relationships between the construction teams involved in a construction project Construction management aims to reduce the performance variability of the construction teams involved in a construction project. Construction management aims to reduce the external interference experienced by the construction teams involved in a construction project.

There are a number of conditions in addition to the basic propositions which need to be met in order to achieve efficient construction. These efficiency conditions relate to construction teams, relationships, construction companies and the characteristics of organizations.

4.13 Construction teams efficiency conditions

Efficient construction teams exhibit a set of common characteristics.

Construction Team Efficiency Conditions

To achieve overall efficiency construction teams need to be competent in the technologies required by the project in which they are involved.

To achieve overall efficiency construction teams need to accept the agreed objectives of the project in which they are involved.

To achieve overall efficiency construction teams need to be motivated to achieve the agreed objectives of the project in which they are involved.

A construction team's competence in any given technology is related to their education and training, the length of their effective experience of that technology and the extent to which they adopt effective new developments in the technology.

In broad terms their education and training provide the essential entry qualifications to begin gaining relevant experience. Practical experience develops competence which inevitably is low initially, improves rapidly with more experience, and reaches a plateau. It is common for competence to eventually deteriorate due to increasing age, boredom with undertaking the same tasks, and gradual improvements in general standards as new entrants find more effective methods of applying the technology. The deterioration can be delayed by: actively monitoring new developments; regular training to test and improve competence; and initiatives which maintain the team's active interest in their work.

This normal pattern of competence depends on reasonably continuous experience. Where this is interrupted, competence falls back roughly in relation to the length of the gap in effective experience of that technology. The previous level of competence is usually regained after a relatively short period of practice following the interruption.

Construction teams have individual objectives. Some of the objectives of the customer and the construction teams involved in a project will be mutually consistent but others will conflict. The conflicting objectives need to be discussed and a consensus agreed. The extent to which construction teams accept the agreed objectives is related to their perception of their involvement in finding the consensus and the extent to which they regard the outcome as advantageous to themselves.

A construction team's motivation is related to their perception of the relationship between their efforts and rewards. Individual construction teams views about efforts and rewards are determined by their individual value systems and the terms of their employment in the project. Construction management needs to be aware of these factors as they make decisions which affect the construction teams.

4.14 Propositions about construction management decisions relating to construction teams

The next group of propositions about construction management decisions are shaped directly by the propositions about construction teams.

Construction Team Propositions
Construction management aims to select construction teams competent in the technologies required by the project in which they are involved. Construction management aims to ensure construction teams accept the agreed objectives of the project in which they are involved. Construction management aims to ensure construction teams are motivated to achieve the agreed objectives of the project in which they are involved.

4.15 Construction team relationships efficiency conditions

Competent teams which accept and are motivated to achieve agreed objectives need conditions that enable them to establish effective relationships if they are to be fully efficient.

Construction Team Efficient Relationships Conditions
To achieve overall efficiency communication between the construction teams involved in a project organization needs to be accurate. To achieve overall efficiency the attainment of accurate communication between the construction teams involved in a project organization needs to be effortless. To achieve overall efficiency the negotiations needed to agree the transactions which bring construction teams into a project organization need to be short and seamless. To achieve overall efficiency the transactions which brought the construction teams into a project organization need to be seen by them as advantageous. To achieve overall efficiency construction teams should not use resources in attempting to improve the terms of the transactions which brought them into a project organization.

These conditions reflect differences between boundary and internal relationships. As described earlier in the chapter, boundary relationships are purely contract-based and involve extensive communication and almost endless negotiations over the terms and conditions of the transactions which bring construction teams into a project organization. In contrast, internal relationships enable construction teams to concentrate on effective actions and so achieve high levels of efficiency. Construction management strategies which encourage interacting teams to achieve accurate communication with minimum effort and resources, and accept the terms of their transactions, can transform boundary relationships into established relationships. When these benefits are repeated over a number of projects, the relationships can become equivalent to internal relationships.

4.16 Propositions about construction management decisions relating to construction team relationships

The propositions about the quality of relationships between construction teams give rise to the following propositions about construction management decisions.

Construction Team Relationship Propositions

Construction management aims to foster accurate communication between the construction teams involved in a project organization.

Construction management aims to minimise the effort required to achieve accurate communication between the construction teams involved in a project organization.

Construction management aims to minimise the length and intensity of negotiations needed to agree the transactions which bring construction teams into a project organization.

Construction management aims to ensure construction teams regard the transactions which brought them into a project organization as advantageous to themselves.

Construction management aims to minimise the resources construction teams devote to improving the terms of the transactions which brought them into a project organization.

4.17 Construction companies' efficiency conditions

Overall efficiency depends on construction companies' understanding of local markets, the skills and knowledge possessed by their construction teams, the quality of the relationships between their construction teams, the quality of support provided for their construction teams, and the extent to which the companies foster relevant innovations.

Construction Company Efficiency Conditions

To achieve overall efficiency construction companies need to be able to satisfy the requirements of local or specialised construction markets in ways which ensure the company's long-term survival.

To achieve overall efficiency construction companies need to provide construction teams with well-developed skills and knowledge which match the requirements of construction projects.

To achieve overall efficiency construction companies need to provide integrated construction teams with established relationships.

To achieve overall efficiency construction companies need to fully support their construction teams.

To achieve overall efficiency construction companies need to foster innovations which match the requirements of construction projects.

4.18 Propositions about construction management decisions relating to construction companies

The propositions about construction companies give rise to the following propositions about construction management decisions.

Construction Company Propositions

Construction management aims to satisfy the requirements of local or specialised construction markets in ways which ensure their company's long-term survival.

Construction management aims to develop construction teams with well-developed skills and knowledge which match the requirements of construction projects.

Construction management aims to develop construction teams integrated by established relationships.

Construction management aims to improve the quality of support provided to construction teams by the construction companies of which they form part.

Construction management aims to foster innovations which match the requirements of construction projects.

These propositions relate to all types of construction companies whatever construction actions they undertake. Thus, it includes those involved in preparing briefs, designs and plans, and undertaking procurement, manufacturing, production and commissioning. It includes those undertaking essential supporting actions to these seven essential actions. It is important that these include the companies which form the supply chains which link manufacturing to construction sites.

4.19 Common organizational characteristics efficiency conditions

The next group of conditions derive from the common characteristics of construction organizations. They deal with their internal actions and the quality of their relationships with each other.

Common Organizational Characteristics Efficiency Conditions
To achieve overall efficiency construction organizations need to use effective information systems. To achieve overall efficiency construction organizations need to consistently establish common values through every aspect of their work. To achieve overall efficiency all the parts of construction organizations need to act in ways which are intended and authorised. To achieve overall efficiency construction organizations need to ensure communications are effective and result in common understandings. To achieve overall efficiency construction organizations need to ensure transactions are agreed with the minimum of effort, accepted as fair by all the parties involved, foster established relationships, and are acted on in the spirit in which they were agreed. To achieve overall efficiency construction organizations need to form established relationships.

4.20 Propositions about construction management decisions relating to common characteristics of construction organizations

The propositions about the common characteristics of construction organizations give rise to the following propositions about construction management decisions.

Common Organizational Characteristics Propositions
Construction management aims to ensure their organization uses effective information systems. Construction management aims to establish values which run consistently through every aspect of their organization's work. Construction management aims to ensure all the parts of their organization act in ways which are intended and authorised. Construction management aims to ensure their organization's communications are effective and result in common understandings. Construction management aims to ensure their organization's transactions are agreed with the minimum of effort, accepted as fair by all the parties involved, foster established relationships, and are acted on in the spirit in which they were agreed. Construction management aims to ensure their organization forms established relationships with other organizations.

These propositions relate to all types of construction organization. They provide guidance to construction managers as they make decisions about their organizations irrespective of its specific characteristics. Thus, they are relevant to construction managers responsible for construction teams, companies, company divisions, project organizations and programme organizations. The practical effects vary depending on the size, particular specialisms, areas of business, internal organization and objectives of individual organizations but the propositions are universally valid. The consequences for practice are described in Chapter 10.

4.21 Double-loop learning condition

The next condition which influences efficiency derives from the concept of double-loop learning and the vital concept of feedback.

Double-Loop Learning Conditions
To achieve day-to-day efficiency construction organizations need to collect, review and act on feedback about the effects of the organization's actions on its objectives. To achieve continuous improvements in efficiency, construction organizations need to collect, review and act on feedback about the effects of established norms and procedures.

4.22 Propositions about construction management decisions relating to double-loop learning

The proposition about double-loop learning gives rise to the following proposition about construction management decisions.

Double-Loop Learning Propositions
Construction management aims to ensure their construction organization collects, reviews and acts on feedback about the effects of the organization's actions on its objectives. Construction management aims to ensure their construction organization collects, reviews and acts on feedback about the effects of established norms and procedures.

This final proposition is important in construction because the traditional project bias of much practice tends to result in much valuable feedback never being collected, reviewed or acted on. Long-term efficiency is absolutely dependant on feedback guiding every part of construction organizations. Every construction team should have excellent feedback about its own performance and be motivated to learn from it and develop more effective ways of working. The same should be true for every project organization, programme organization, company division and company in the construction industry.

4.23 Construction efficiency

The fundamental construction management theorem states the purpose of construction management is to enable construction to be undertaken efficiently. This raises the issue of how efficiency should be measured. In general terms efficiency is related to the extent to which agreed objectives are achieved. In practical terms it means minimising the amount of waste resulting from complexity and external interference which prevents agreed objectives being achieved.

These definitions emphasise the importance of agreed objectives in shaping construction management strategies. This important influence is taken into account in the descriptions of the major construction management approaches described in Chapter 5 to 9.

The concept of agreed objectives recognises that construction is undertaken by construction teams which simultaneously form part of construction project organizations and construction companies. Some construction project organizations are established within construction programme organizations. Some construction companies are organized in divisions. It follows that all of these organizations may have objectives for individual construction projects.

Customers and construction organizations have different objectives which are likely to change over time. It is therefore inevitable that the objectives for any construction project result from interactions between the organizations involved. This is why the focus of the theory of construction management is **agreed objectives** which are the set of aims which motivate organizations responsible for undertaking construction. They provide the last of the basic concepts used in the theory of construction management.

The key relationships between all the basic concepts are illustrated in Figure 4.5.

4.24 Inherent difficulty indicators

Figure 4.5 makes clear the central role of inherent difficulty in shaping construction management strategies. This is a complex concept and the practical and academic issues involved in measuring inherent difficulty are discussed in Chapter 11. This makes clear that robust measures require considerable time and effort in setting up and running sophisticated simulations of individual projects. It can reasonably be expected that future research will make it possible to produce robust measures of inherent difficulty rapidly and economically.

It is nevertheless possible to describe the basic concepts used in the theory of construction management in mathematical terms to provide effective measures of features of construction which have a crucial impact on construction management decisions. This provides inherent difficulty indicators (IDIs) which provide a practical basis for selecting construction management strategies. The use of IDIs is illustrated in Chapters 5 to 9 where they are used to demonstrate the effect of the major construction management strategies.

Established relationships

The total number of construction teams involved in a construction project is N. The total number of possible relationships between teams within the project organization is then:

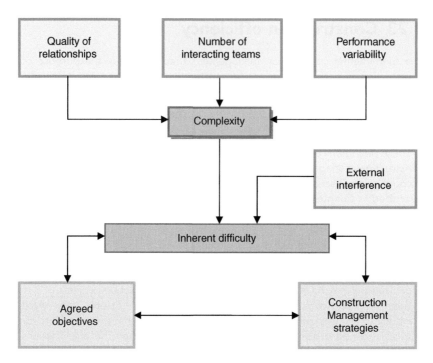

Figure 4.5 Basic concepts in the theory of construction management

$$R = \frac{N(N-1)}{2} \qquad [4.11]$$

However, some construction teams complete their involvement in the project long before other teams even begin their work and project complexity depends only on relationships between directly interacting teams so we can disregard inconsequential relationships (k). For example, a team involved in earthworks has no direct interaction with a team involved in roof works so we may disregard such relationships. Of course, this is not to say a delay caused by earthworks would not impact later actions but at this stage we need to identify direct interactions between project teams as they are a major complexity indicator.

The number of interacting teams at time t there are $N(t)$ interacting teams and the number of relationships is:

$$n(t) = \frac{N(t)(N(t)-1)}{2} \qquad [4.12]$$

As the number of teams increases linearly, the number of relationships increases to the power of two, showing how important the number of involved teams is in the attempt to reduce complexity. Many relationships are new as teams have never worked together before and projects rarely provide sufficient time to build effective relationships. An important consequence in many project organizations is teams tend to resort to their contractual obligations and only slowly, if at all, build an effective pattern of working together.

There may be relationships identified in equations 4.11 and 4.12 that were established before the project started. They are a culmination of relationships between construction teams on a number of previous projects. Such relationships can be designated as established relationships between interacting construction teams. So, if the total number of established relationships is n^*, then it is possible to determine an indicator of established relationships in the following way:

$$E_R = \frac{n^*}{\frac{N(N-1)}{2} - k} = \frac{2n^*}{N(N-1) - 2k}$$

[4.13]

Relationship fluctuation

While E_R might be the same for several projects, this may hide important differences in the impact of established relationships. In some projects the number of established relationships will be consistent for all time intervals. However, in other projects there may be time intervals where none of the teams have worked together before and time intervals where all the teams have established relationships from previous projects. To measure these differences we can begin by taking account of the number of established relationships between interacting teams at time t only, which yields:

$$E_R(t) = \frac{n^*(t)}{\frac{N(t)(N(t)-1)}{2}} = \frac{2n^*(t)}{N(t)(N(t)-1)}$$

[4.14]

where $N(t)$ represents the number of directly interacting teams at time t.

The above indicator may have values between 0 and 1 corresponding to how far the project organization is from the ideal configuration in which all teams have been working together before. If the indicator equals one then all interacting teams have worked together before, if it equals zero then none of the teams have worked together before and all relationships in the project organization are new.

At different times $E_R(t)$ will assume different values due to a change in the number of teams working together at the same time and different patterns of established relationships. This change can be expressed in terms of the number of independent time intervals v, where each time interval is a period which has a consistent configuration. In total there can be a maximum of $v_{max} = 2N - 1$ time intervals with different configurations of teams. A high level of fluctuations in $E_R(t)$ indicate a project in which construction management has to deal with many changes in the configuration of teams. The standard deviation from the mean provides a measure of fluctuations in the number of teams and the established relationships over the separate time intervals throughout the project. The maximum possible value of the standard deviation for values between zero and one is 0.707 (i.e. two intervals with $E_R(t)$ values of zero and one respectively), which is less than one so the relationship fluctuation indicator needs to be divided by 0.707 to make it consistent with the other indicators:

$$F_E = 1 - \frac{\sqrt{\frac{1}{v}\sum_v (E_{Rv} - \mu)^2}}{0.707}$$

[4.15]

where μ is the mean value, E_{Rv} is $E_R(t)$ in interval v and 0.707 is a unit correction factor.

Relationship quality

Not all established relationships are of equal quality. Some construction teams may have worked together before but only on a single project while others have spent years working on a number of different projects all over the world. Therefore the time interacting construction teams have worked together before provides a relationship quality indicator. Let us therefore assume that on the current project two teams in the i-th relationship work together t_i days. Furthermore, let us assume that these two teams have worked together on previous projects for a total duration of T_i days. In addition, weights (w_i) can be used to recognise that some relationships may be critically but others only marginally important. Taking these assumptions into consideration, we can determine a relationship quality factor for each relationship in the following way:

$$Q_{Ri} = \left(1 - \frac{t_i}{T_i + t_i}\right) w_i \qquad i = 1, \ldots, n(t) \qquad [4.16]$$

The total established relationship quality factor could then be calculated in the following way:

$$Q_R = \frac{\sum_i Q_{Ri}}{R - k} \qquad [4.17]$$

Relationship configuration

Equations 4.12 and 4.14 indicate that at different times we may have several teams working on overlapping activities, which means there are in total $v_{max} = 2N - 1$ intervals with different team constitutions and consequently different interactions. Relationship configuration indicator C_R measures these patterns in order to indicate the inherent difficulty of a project. It is calculated in the following way:

$$C_R = \left(1 - \frac{v}{v_{max}}\right)\alpha + \left(1 - \frac{\bar{t}}{T_p}\right)\beta + \left(1 - \frac{N(t)_{max}}{N}\right)\gamma \qquad \alpha + \beta + \gamma = 1 \quad [4.18]$$

where:

- v represents the number of independent time intervals in a project with different constitution of interactions;
- v_{max} represents the number of all possible time intervals in a project;
- $\bar{t}$ represents the average duration of time intervals in a project which involve two or more teams;
- T_p represents project duration;
- $N(t)_{max}$ represents the maximum number of directly interacting teams per time interval throughout the project.
- α, β, γ factors denoting relative importance of the number of time intervals, length of time intervals and maximum number of simultaneously interacting teams

A worked example of the relationship configuration indicator is given below. It illustrates how equation 4.18 comprises three major elements. The first bracket represents the influence of the number of time intervals with a different pattern of interactions. The more time intervals there are the more changes to the construction project organization have to be managed. For example, teams that have established common ground and effective relationships may at some later time be joined by a team with which similarly effective relationships are yet to be established. Figure 4.6b shows a case where a new team joins or leaves the existing configuration every two days. This is likely to require considerable effort from construction management to ensure a smooth transition and even reasonably effective relationships. On the other hand, a low number of time intervals, as in the hypothetical case shown in Figure 4.6a, reduces the number of such transitions (i.e. new teams joining the existing teams or some existing teams leaving the project). A low number of transitions however, requires less management effort to deal with the effects of transitions allowing construction managers to focus solely on ongoing interactions. The second bracket represents the influence of the length of time intervals. Short time intervals normally lead to little or limited interaction but longer time intervals indicate much higher required intensity of interactions (i.e. the longer the time intervals in relation to project duration, the more complex the configuration). Even for teams that have worked together on past projects, longer time intervals lead to a higher probability that during intensive interactions something could go wrong. The third bracket represents the influence of the number of interacting teams. In some cases there will be only a few directly interacting teams in each time interval out of the total number of teams on the project but in some other cases the number of directly interacting teams will be much higher.

Example – Relationship Configuration Indicator

Figures 4.6a, 4.6b and 4.6c show three different configurations of four actions. The first configuration comprises a single time interval with a total length of 8 days; the second configuration comprises the maximum number of 7 time intervals with an average duration of 2 days; and the third configuration comprises 3 time intervals with an average duration of 3.3 days.

Using equation 4.18 we can calculate C_R for all three configurations (to simplify we assume that $\alpha = \beta = \gamma = 0.33$):

First configuration: $C_R = \frac{1}{3}\left(\left(1 - \frac{1}{7}\right) + \left(1 - \frac{8}{8}\right) + \left(1 - \frac{4}{4}\right)\right) = 0.29$

Second configuration: $C_R = \frac{1}{3}\left(\left(1 - \frac{7}{7}\right) + \left(1 - \frac{2}{14}\right) + \left(1 - \frac{4}{4}\right)\right) = 0.29$

Third configuration: $C_R = \frac{1}{3}\left(\left(1 - \frac{3}{7}\right) + \left(1 - \frac{3.3}{10}\right) + \left(1 - \frac{4}{4}\right)\right) = 0.41$

The third configuration is the least inherently difficult in terms of its relationship configuration indicator. There are only three distinct time intervals as opposed to seven in the second configuration and there are only two directly interacting teams in the first and third interval, which in total represents 40 per cent of the project. The first and second configurations are equivalent. While there is only one time interval in the first configuration, all teams have to interact throughout the duration of the project indicating a high intensity of interactions for the whole duration of the project. The second configuration comprises the highest number of time intervals indicating much higher interaction dynamics but time intervals are short, which compensates for the increased difficulty due to the number of distinct time intervals.

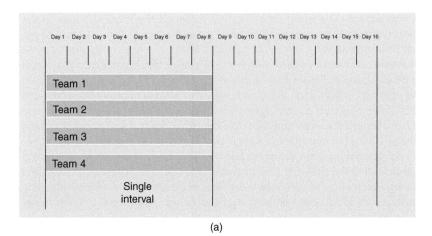

(a)

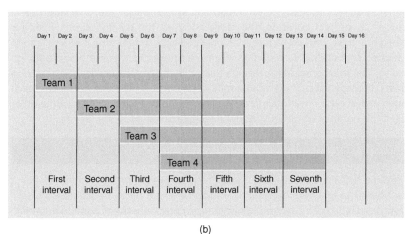

(b)

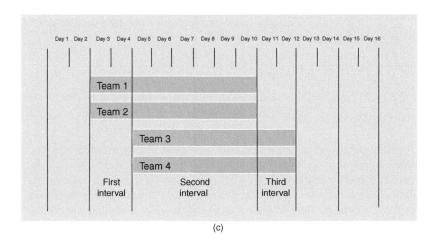

(c)

Figure 4.6 (a) The First Configuration Shows Four Actions Running in Parallel Over a Single Interval ($v = 1$); Project Duration is 8 Days. (b) The Second Configuration Shows Four Partially Overlapping Actions with $v_{max} = 7$ Independent Intervals with Different Teams Interacting; Project Duration is 14 Days. (c) The Third Configuration Shows Four Partially Overlapping Actions with $v = 3$ Independent Intervals with Different Constitution of Interactions; Project Duration is 10 Days.

Performance variability

In practice team performance may not be consistent from one project to another. In fact, the performance of some teams may be substantially more inconsistent than that of other teams even within the same project. Because different teams perform different actions, it is not possible to directly compare how they perform but it is possible to measure the consistency of each team's performance. Widely inconsistent performance would identify a team as unreliable and therefore likely to create difficulties for construction management. An individual team's past performance record would reveal the likelihood of the team completing their work on time. The distribution function generated by a team's performance record could be bell shaped, it could be skewed in favour of exceptional or poor performance, it could have high peaks with fat tails and so on but most importantly, it would show the percentage of actions the team have completed early and on time. To determine a performance variability indicator, let's assume for the j-th team there are n_{oj} occurrences of work being completed on time, n_{ej} occurrences of work being completed early and n_{lj} occurrences of work being completed late. The performance variability indicator could then be calculated as a percentage of occurrences completed early or on time in the following way:

$$R_{Pj} = \frac{n_{oj} + n_{ej}}{n_{oj} + n_{ej} + n_{lj}} \qquad j = 1, \ldots, N \qquad [4.19]$$

This hugely simplified equation does not take account of the properties of possible distributions such as variance, skewness and kurtosis but it is sufficiently robust to provide an indicator of a particular team's performance variability. However, projects involve a number of teams which requires an aggregate indicator of performance variability. The aggregate performance variability indicator could be a simple average of all team performance variability indicators in the following form:

$$R_P = \sum_j \frac{R_{Pj}}{N} C_j \qquad 0 \le c_j \le 1 \qquad [4.20]$$

where c_j represents a performance variability impact indicator for team j.

In some instances performance variability of one team may have far greater overall impact on overall efficiency than that of all other teams. The impact indicator c predominantly depends on the core construction technology. For example, high performance variability by a steelwork construction team working in a steel-frame construction project would have greater impact on overall efficiency than the equivalent performance variability of other teams on the project. In this example steelwork is the lead technology and as such sets the tempo of work for other teams. Consequently inadequate performance by the steelwork construction team would have the greatest effect on overall performance. In a similar way performance variability by bricklaying teams would have the greatest impact on projects with traditional brick structures, performance variability by concreting teams would have the greatest impact in concrete-frame projects, and so on. The impact indicator may assume any value between zero and one and can be determined from historical records of the impact of performance variability by various construction teams. For instance, the impact of performance variability by bricklaying teams may be greater in projects with traditional brick structures than that of the steelwork construction teams in steel-frame projects.

The distribution of variability figures could reveal further properties. For example, in one case the variability figures for all teams may be close to the mean with a very low standard deviation but in other cases the deviation from the mean may be far greater and this is likely to increase the inherent difficulty faced by construction management.

External interference

External interference may cause delays to a construction project. The cause may be any one of a multitude of factors outside the control of the project organization. These include extreme weather events, public demonstrations and sudden political changes. These external perturbations are rarely predictable and often cause costly delays to construction. While some may only last a few hours, other events can bring construction to a halt for several days, weeks or even months. No matter how efficient a project organization is and how many established relationships exist, prolonged perturbations inevitably have a considerable effect on the overall efficiency of construction.

Forming judgements about the possible impacts of external interference provide a major challenge for construction management because it is so unpredictable. Formal risk registers can provide a detailed picture of past events, their frequency and impact, and can be useful in making judgements about the likelihood and duration of potential perturbations. Risk management is an essential part of modern construction management and is supported by an extensive literature.

The following equation assumes records of external interference on similar past projects are available. The equation calculates the total duration of all perturbations as a proportion of total project duration to provide a good indication of the impact on the inherent difficulty. This can be expressed as an external interference indicator in the following way:

$$ I = \frac{\sum_j \left(\left(1 - \frac{t_j}{T}\right) a_j \right)}{m} \qquad [4.21] $$

where m is the total number of perturbations, t_j is the duration of j-th perturbation, a_j is the indicator of interference magnitude representing the relative impact of j-th perturbation, and T is the total duration of a project.

4.25 IDIs in practice

The IDIs: *established relationships, relationship fluctuation, relationship quality, relationship configuration, performance variability* and *external interference* provide a basis for assessing the inherent difficulty of a construction project which can be calculated sufficiently early in a project to provide guidance in the choice of construction management strategy. If all the indicators are close to one there is a high probability the construction project will be successful. The following example illustrates how the equations are used in calculating the IDIs.

Chapters 5 to 9 use the IDIs to demonstrate the effect of the construction management strategies described. Chapter 10 describes the practical implications and Chapter 11 explains how the equations provide a basis for research into construction management and as part of this more highly developed ways of measuring inherent difficulty are discussed.

Example – Inherent Difficulty

We will look at an example of a minor project comprising six actions and running over 16 days as depicted in the simplified Gantt chart in Figure 4.7.

In this case, from equation 4.11 there are $R = 15$ possible relationships between teams working on this project. Because Team 1 complete their work before Teams 3, 4 and 5 even start we will disregard the $k = 3$ possible relationships between Team 1 and Teams 3, 4 and 5, yielding a total of $R-k = 12$ interactions that do in fact occur.

In addition and according to equation 4:12 there are several interacting teams at different time intervals. On days 1, 2 and 3 there is only Team 1 and the Design Team $[n(t)_1 = 1]$, these are joined by Team 2 from day 4 onwards $[n(t)_2 = 3]$, we have only Team 2 and the Design Team on Day 7 $[n(t)_3 = 1]$, Teams 2, 3, 4 and the Design Team on days 8, 9, 10 and 11 $[n(t)_4 = 6]$, we have Teams 2, 3, 4, 5 and the Design Team on day 12 $[n(t)_5 = 10]$, Teams 4, 5 and the Design Team on day 13 and 14 $[n(t)_6 = 3]$, and finally Team 5 and the Design Team on days 15 and 16 $[n(t)_7 = 1]$.

The maximum number of relationships appears on day 12. Apart from Team 1 all other teams are present on that day so there are 10 different relationships. Next we need to identify how many of these 10 relationships are established relationships. Let's assume Teams 2 and 3 worked together on several projects in the past, and Teams 3 and 5 also worked together on several past projects. This would give an Established Relationships Indicator: $E_R = 2/12 = 0.17$ (equation 4.13). This is a very low value which is then further confirmed by equation 4.14 which generates the following time dependent indicators of established relationships: $E_R(t)_1 = 0$, $E_R(t)_2 = 0$, $E_R(t)_3 = 0$, $E_R(t)_4 = 1/6 = 0.16$, $E_R(t)_5 = 2/10 = 0.2$, $E_R(t)_6 = 0$ and $E_R(t)_7 = 0$.

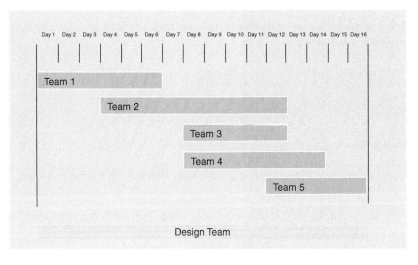

Figure 4.7 Minor project comprising 6 actions with a total of 6 teams

There are only two time intervals where some teams have some prior experience of working together before and even in these cases the values suggest the prior experience will provide little benefit.

The relationship fluctuation indicator F_E is calculated using equation 4.15 and gives a value of 0.88. This reflects a reasonably consistent influence of established relationships throughout the project but this is not particularly significant because on this project there are few established relationships.

However, to really establish the benefits of established relationships (Teams 2 & 3 and Teams 3 & 5), we have to look at how long they are working together on this project and also know how much time Teams 2 & 3 and Teams 3 & 5 have spent working together on past projects. Taking into account equation 4.16 we assume the following:

- Relationship between Teams 2 & 3 is more important than that between Teams 3 & 5 because they have to work together 5-times longer on this project, so the weighting indicators are 0.8 and 1 respectively (i.e. the lower the weighting indicator the more important the relationship).
- Teams 2 & 3 worked together on two past projects with a total duration of 50 days;
- Teams 3 & 5 worked together on 4 past projects with a total duration of 120 days.

The Relationship Quality Indicator for each relationship would thus be: $Q_{R(2\&3)} = 0.72$ and $Q_{R(3\&5)} = 0.99$ but the total Relationship Quality Indicator (equation 4.17) is: $Q_R = (0.72 + 0.99)/12 = 0.14$. Although we have high values for each individual Relationship Quality Indicator, the total Relationship Quality Indicator is very low because there are only two established relationships in this project out of possible twelve.

According to Eq. 4.18 we can calculate the relationships configuration indicator C_R (to simplify we assume that $\alpha = \beta = \gamma = 0.33$):

$$C_R = \frac{1}{3}\left(\left(1 - \frac{7}{11}\right) + \left(1 - \frac{2.28}{16}\right) + \left(1 - \frac{5}{6}\right)\right) = 0.46$$

The past performance of the five teams over their immediately previous 10 projects is as follows:

Team 1: $n_{e1} = 1$, $n_{o1} = 2$, $n_{I1} = 7$; $c_1 = 0.8$
Team 2: $n_{e2} = 3$, $n_{o2} = 3$, $n_{I2} = 4$; $c_2 = 0.8$
Team 3: $n_{e3} = 2$, $n_{o3} = 3$, $n_{I3} = 5$; $c_3 = 1$
Team 4: $n_{e4} = 0$, $n_{o4} = 3$, $n_{I4} = 7$; $c_4 = 1$
Team 5: $n_{e5} = 6$, $n_{o5} = 3$, $n_{I5} = 1$; $c_5 = 0.8$
Design Team: $n_{e6} = 2$, $n_{o6} = 4$, $n_{I6} = 4$; $c_6 = 0.8$

Using equation 4.19 we obtain the following individual team Performance Variability Indicators:

$$R_{p1} = 0.3, \ R_{p2} = 0.6, \ R_{p3} = 0.5, \ R_{p4} = 0.3, \ R_{p5} = 0.9, \ R_{p6} = 0.6$$

The total Performance Variability Indicator (equation 4.20) is:

$$R_p = (0.24 + 0.48 + 0.5 + 0.3 + 0.72 + 0.48)/6 = 0.44$$

This is of course the average performance variability showing the teams managed to complete their work within the agreed time on 44 per cent of their recent projects.

To establish the External Interference Indicator, we will assume historical records for the particular geographical region for this type of project show an average of one weather-related and one Health & Safety (H&S) perturbation lasting approximately one day each. The relative impact of the weather-related and H&S perturbations is represented by $a_1 = 0.8$ and $a_2 = 0.6$ respectively. Using equation 4.21 to calculate the External Interference Indicator gives:

$$I = \frac{\left(1 - \frac{1}{16}\right)0.8 + \left(1 - \frac{1}{16}\right)0.6}{2} = 0.66$$

Although both external interference events have only limited impact on the project the above indicator shows even very short perturbations can have a negative effect on construction projects.

Overall, we have obtained the following major indicators:

Established Relationships: $E_R = 0.17$ ($E_{R,max} = 1.0$)

Relationship Fluctuation: $F_E = 0.88$ ($F_{E,max} = 1.0$)

Relationship Quality: $Q_R = 0.14$ ($Q_{R,max} = 1.0$)

Relationship Configuration: $C_R = 0.46$ ($C_{R,max} = 1.0$)

Performance Variability: $R_p = 0.44$ ($R_{p,max} = 1.0$)

External Interference: $I = 0.66$ ($I_{max} = 1.0$)

The low value of most of the indicators suggests the project is inherently difficult and there is a high probability it will not be completed as planned. The teams have a relatively poor performance record and although there are two established relationships which benefit only a small proportion of the interactions. Given the likelihood of some external interference, it is clear the project organization will have a difficult job to bring the project to completion exactly as planned unless they adopt a construction management strategy which reduces the inherent difficulty.

4.26 Size of construction projects

The IDIs provide guidance on selecting effective construction management strategies for individual projects. Construction managers can compare the IDIs for a proposed project with the strategies and performance of similar projects which have similar IDIs.

The IDIs directly measure the significant characteristics of construction projects with one important exception. The exception is the size of the projects which should be taken into account in making effective comparisons. In practice construction companies, divisions of companies and individual construction teams specialise in projects of a particular size. This is because projects of markedly different sizes have distinct characteristics. Minor projects have their own characteristics which are not shared by larger projects. The time and resources available to undertake mega-projects is fundamentally different from smaller projects. For example, it is common for mega projects and even large projects to be divided into project sized subprojects each of which have a substantially independent construction project organization. For example, low-rise housing projects are subdivided by experienced construction managers into independent projects comprising between 25 and 30 individual houses. Similarly distinct characteristics are found in each size of project.

The most effective way of taking account of these effects is to classify sets of IDIs in terms of project size. The most practical unit for measuring the size of construction projects is team-days. This is because construction management is concerned with the selection and organization of teams; days are the smallest unit normally used in planning construction actions. More than this construction teams tend to think and work in terms of days.

The classification scheme shown in Figure 4.8 provides five distinct categories which reflect normal construction terminology. The boundaries between the categories are vague to indicate that near the boundary the classification of individual projects should be based on experienced judgement. In this way the classification provides a practical and robust basis for using IDI data effectively.

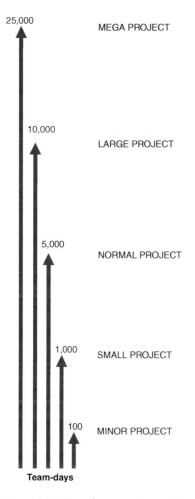

25,000   MEGA PROJECT

10,000   LARGE PROJECT

5,000   NORMAL PROJECT

1,000   SMALL PROJECT

100   MINOR PROJECT

Team-days

Figure 4.8 Size of construction projects in team-days

4.27 Using the theory of construction management

The construction management propositions provide a coherent theory to guide the choice of construction management strategies. The IDIs are designed to help identify relevant previous projects which can be used to check the likely effect of selecting any particular construction management strategy. The theory and the IDIs are used in Chapters 5 to 9 to help explain the most widely used and important construction management strategies. The practical implications are described in Chapter 10. Proposals for testing and developing both the theory and the IDIs are described in Chapter 11.

Exercise – Inherent Difficulty Indicators

Consider the minor project illustrated in Figure 4.9 comprising 6 different teams. Based on the past records, the following is known about the involved teams:

1. Teams 1 & 2 worked together on several past projects with a total duration of 200 days and a weighting factor $w_{1-2} = 1$;
2. Teams 2 & 3 worked together on several past projects with a total duration of 100 days and a weighting factor $w_{2-3} = 0.7$;
3. Teams 2 & 4 worked together on several past projects with a total duration of 300 days and a weighting factor $w_{2-4} = 0.9$;
4. Teams 3 & 4 worked together on several past projects with a total duration of 50 days and a weighting factor $w_{3-4} = 0.75$;
5. Teams 3 & 5 worked together on several past projects with a total duration of 70 days and a weighting factor $w_{3-5} = 1$;
6. Teams 5 & 6 worked together on several past projects with a total duration of 120 days and a weighting factor $w_{5-6} = 0.6$.
7. Teams past performance:
   Team 1: $n_{e1} = 1$, $n_{01} = 3$, $n_{I1} = 6$;
   Team 2: $n_{e2} = 2$, $n_{02} = 4$, $n_{I2} = 4$;
   Team 3: $n_{e3} = 3$, $n_{03} = 5$, $n_{I3} = 2$;
   Team 4: $n_{e4} = 3$, $n_{04} = 3$, $n_{I4} = 4$;
   Team 5: $n_{e5} = 1$, $n_{05} = 5$, $n_{I5} = 4$;
   Team 6: $n_{e5} = 3$, $n_{05} = 6$, $n_{I5} = 1$;
   Design Team: $n_{e6} = 1$, $n_{06} = 4$, $n_{I6} = 5$.
8. Past records show on average projects of this type are affected by one 2-day long weather-related perturbation and one health and safety related perturbation lasting one day. The relative impact of the

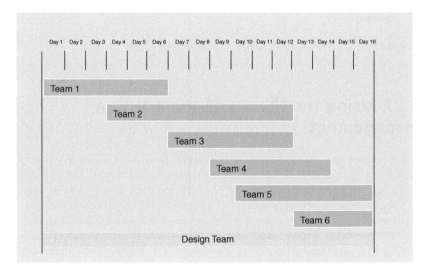

Figure 4.9 Minor project comprising 6 actions with a total of 7 teams

weather-related and health and safety related perturbations is represented by $a_1 = 0.6$ and $a_2 = 0.7$ respectively.

a. Calculate the following indicators:

   How many relationships (k) could be disregarded?

   Calculate the number of relationships of interactions at different time intervals N(t).

   Calculate the total E_R and $E_R(t)$.

   Calculate the individual and the total Relationship Quality Indicator: Q_R.

   Calculate the Relationship Configuration Indicator: C_R.

   Calculate the individual and the total Performance Variability Indicator: R_p.

b. Use the same example but change the weighting factors, durations and variability figures to obtain a new set of indicators. What differences can you observe? What does that tell you about the changes you have made?

c. What additional changes would need to be satisfied to achieve efficient construction? Hint: read the construction management propositions.

Further Reading

The following publications are the source of ideas used in this chapter and provide further information for readers.

Hall, P. (1980) *Great Planning Disasters*. Weidenfeld & Nicolson. This classic text provides a series of examples of the way major projects, including construction projects, can go badly wrong. The lessons identified are taken into account in this chapter.

Ward, S. and Chapman, C. (2008) Stakeholders and uncertainty management in projects. *Construction Management and Economics*, 26(6), 563–77. This paper describes a structured approach to managing uncertainty in projects and provides a clear overview of the importance of risk management.

Winch, G. M. (2010) *Managing Construction Projects*. Wiley-Blackwell. The second edition of this influential book provides a mass of practical and theoretical background for the construction management concepts described in this chapter. Each chapter can be used as a standalone learning platform with a separate reference list, further reading and a case study. Its strength is in breadth of coverage of project management whether it is about managing the schedule, leadership or project information flow.

Chapter Five
Traditional Construction

5.1 Introduction

This chapter describes traditional construction which is defined in Chapter 4 as the locally established actions of construction project organizations. Traditional construction in most countries has developed gradually over time. The chapter explains why it has changed and how modern developments in construction leave it suitable for an increasingly narrow range of projects. Traditional construction does however continue to provide an appropriate approach for some construction projects. It is also an essential starting point for understanding why other construction management approaches have emerged.

The chapter includes an example of a project using a modern form of traditional construction which demonstrates the use of the Inherent Difficulty Indicators described in Chapter 4. The IDIs serve to illustrate the challenges faced by projects using traditional construction when designs include technologies outside the established competence of local construction organizations. Also the IDIs generated by the example provide a datum which is used to illustrate the benefits of the more developed construction management approaches described in Chapters 6 to 9.

5.2 Fundamental traditional construction

Traditional construction developed over centuries in response to new construction technologies and new demands by construction customers. The simplest and most fundamental traditional construction approach developed from actions involved in producing new buildings.

For centuries construction was limited to a few craft-based technologies. They fitted together in ways which were well understood by everyone involved in construction. The traditional technologies normally included masonry, bricklaying, carpentry, joinery, various roof coverings, basic plumbing, plastering and painting. These construction technologies were learnt by craftsmen through many years of apprenticeship. The best craftsmen progressed to become master craftsmen. Each master craftsman organized a small team of craftsmen, apprentices and labourers, took the lead in agreeing work with customers, and organized the team's activities.

The only explicit coordination mechanism was the design resulting from discussions between the customer and the master craftsmen responsible for

Construction Management Strategies: A Theory of Construction Management,
First Edition. Milan Radosavljevic and John Bennett.
© 2012 John Wiley & Sons, Ltd. Published 2012 by John Wiley & Sons, Ltd.

the dominant technology. In many major building projects this was the master mason but master bricklayers or carpenters took the lead when their work provided the main structural elements of the building. Designs described the overall building in terms of the arrangement of spaces and the general appearance. All the details were left to established craft practice guided by discussions on site as and when any problems arose.

Production was greatly assisted by the inherent flexibility of traditional technologies. Minor errors were accommodated by cutting and fitting or modifying subsequent work. This was part of craft practice and required no interaction between construction teams.

The essential simplicity of construction made it practical and efficient for construction customers to deal directly with master craftsmen. Conventions grew up to control the contracts between customers and craft teams which allowed construction to be a straightforward process in which customer and master craftsman agreed what should be constructed as the projects pro-gressed. Normally the customer bought the materials and paid the workers an agreed daily rate.

Some ambitious customers employed architects to design beautiful build-ings. In the main, architects were educated gentlemen who had a local reputation for good taste. They understood the principles of classical design and produced drawings to guide the craftsmen. Their contribution helped improve the appearance of the outputs produced by fundamental traditional construction but did not alter its straightforward approach.

Even today, the smallest and most straightforward projects use the funda-mental approach. For example, individual house owners employ craftsmen directly to undertake repairs or decorations. They discuss what is required, agree the price and time, and then the work is carried out and paid for on completion.

5.3 Project organization

The fundamental approach to traditional construction gave rise to a simple project organization. This was shaped by the pattern of boundary relationships set up by the customer as they employed the craft teams needed to produce the new facility. Sometimes customers who employed architects would require them to negotiate agreements with the master craftsmen. The resulting project organizations are illustrated in Figure 5.1.

Once the master craftsmen had agreed their work with the customer or architect, the resulting project organization was coordinated largely on the basis of established craft practice. Each team of craftsmen completed their work in the manner expected by the next team in the construction sequence. Provided all progressed well, each craft carried out their work without the need for any interaction between the master craftsmen. Only if problems arose was there any need for the teams to interact. When this happened, the formal boundary relationships between the customer and the master craftsman or craftsmen involved in the problem situation were normally sufficient to find a solution.

5.4 Strengths and weaknesses

The fundamental approach to traditional construction was broadly successful. It produced many fine buildings and other structures which are now widely admired.

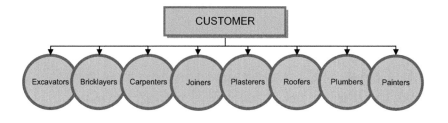

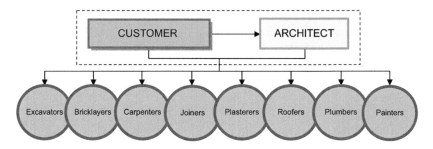

Figure 5.1 Fundamental Traditional Construction Project Organisations.

This basic form of traditional construction tended to be slow, the overall costs emerged as the work progressed which could and often did lead to disputes, quality was generally satisfactory and the resulting new facilities by and large provided what customers required.

Historically a distinct strength of traditional construction was that it was localised with master craftsmen operating in close proximity to their homes. As a consequence, the majority of craftsmen working on local projects knew each other very well because they had worked together on similar projects in the past. They understood each other's work sequence needs so agreeing the steps to completion was relatively straightforward. This basic form of traditional construction has survived to the present day in most parts of the world. Even in the most industrialised countries basic traditional construction still serves a few niche markets which, for example, usually include relatively simple individual house projects.

However, throughout history projects tended to go wrong if the customer or their architect encouraged craftsmen to go beyond the limits of their technologies. Extreme illustrations are provided by ambitious buildings which collapsed during construction or subsequent normal use. On some major projects, this happened several times before a successful approach was found. Indeed some projects were abandoned altogether after repeated failures to find a way of constructing what the customer wanted. Historical examples include Beauvais Cathedral where the vaulting over the choir collapsed in 1284 and the 153 metre high central tower collapsed in 1573; Ely Cathedral where the octagonal tower collapsed in 1322 and the north-west transept collapsed during the fifteenth century and was never rebuilt; and the wooden Rialto Bridge in Venice collapsed in 1444 and again 1524 when it was replaced by the present stone structure.

In the main, fundamental traditional construction suited the conservative, ordered societies it served for many centuries. In general, each generation of craftsmen learnt through a long period of apprenticeship and on the job training to work in the same manner as their predecessors.

5.5 Construction management propositions and fundamental traditional construction

Fundamental traditional construction provided a basic response to the essential nature of construction. It achieved the limited level of efficiency described in the preceding section by restricting designs to well established technologies which required a relatively small number of construction teams and enabled them to undertake their activities largely independently of each other. This resulted in projects which by the standards of modern construction were very simple.

Fundamental traditional construction did not protect projects from variability or interference. However the slow, sequential approach minimised the effects and so most construction teams were not faced with significant uncertainties. Construction took place at an inherently slow, steady pace in generally benign environments. Wars, severe weather, customers losing their fortune or similar major disturbances brought construction to a halt but these events were the exception. More generally fundamental traditional construction faced low levels of inherent difficulty and so could rely on the basic coordination provided by an outline design.

There were however changes emerging, slowly at first and then more rapidly, which revolutionised this straight forward way of undertaking construction.

5.6 New technologies

A major force for change came from new construction technologies. They developed slowly over centuries. For example, glass became more common allowing weather-tight windows to be constructed. Chimneys were developed allowing relatively smoke-free heating and cooking inside buildings. As a result of these and other new technologies, buildings gradually became more comfortable and convenient for their occupants.

Since the early years of the eighteenth century, other important developments were centred on transport systems, including canals, railways and then roads, which heralded the industrial revolution. These transport developments resulted in the emergence of distinct sectors of the construction industry based on ownership of the knowledge and skills needed to construct the new infrastructures.

The nineteenth and twentieth centuries produced an explosion in the number of distinct construction technologies as a direct result or in response to the industrial revolution. Many of these were based on recently discovered engineering principles and new materials. Each new development led to more sophisticated constructed facilities. For example, structural steelwork allowed buildings to be much higher than ever before. The need for vertical access led to the development of lifts and elevators.

From the middle of the nineteenth century new services were developed to make building more convenient and comfortable. Pipes, cables and ducts carried services to all parts of even the largest and highest of buildings. Plant rooms were included in individual buildings which resembled small factories processing electricity, heat, refrigeration, water, gas and other physical phenomena and materials.

Distinct sectors of construction grew up to support these major developments. Each was characterised by construction companies specialising in the research, development, design, manufacture, production and commissioning of distinct technologies. Modern buildings typically require the work of hundreds of companies providing specialised knowledge and skills. The resulting increase in the complexity of construction made design a highly specialised activity. It became essential for customers wanting a new building to employ an architect. Similarly customers wanting some new physical or services infrastructure or a new industrial facility needed to employ design engineers.

As these primary design roles became established, architects and engineers set up professional bodies to guide the development of their knowledge and skills and foster their role within a newly emerging approach to traditional construction. This led to university degree courses in engineering in the nineteenth century and in architecture in the first years of the twentieth century.

Even with the benefit of support from their professional bodies and university courses, it became impossible for individual architects or engineers to comprehend the full implications of the ever increasing number of technologies available. Many designers tried to simplify the challenges posed by the technologically rich environment by specialising in a distinct category of new facility. Even with this advantage, architects in particular needed to work with specialised design consultants and companies on all except the simplest of new facilities. Some designers avoided the detailed complexity created by new technologies by concentrating on a broad strategic role in design. These super-designers often did little more than agree an overall design concept with the customer and were content to leave colleagues to organize the large design teams which turned their broad visions into practical designs.

The new technologies inevitably also changed the manufacturing, production and commissioning activities needed to produce a new facility. As traditional craftsmen were joined by an ever increasing number and variety of specialised construction companies, new facilities became ever more complex. Designs often required novel manufacturing activities and unique combinations of production activities. This inevitably meant construction could not rely on established sequences of activities. Problems arose on site. Engineered components in particular lacked the inherent flexibility of traditional technologies and so could not accommodate minor errors in preceding work. Internal services required distinct technologies to be closely fitted together in complex ways. As a result, production involved considerable interaction between construction teams.

In terms of the theory of construction management, the development of new technologies increased the complexity faced by construction teams. This was caused by the increase in the number of teams, the increased intensity of the interactions between them, and more variable performance as manufacturing, production and commissioning teams struggled with novel designs incorporating ever more technologies.

5.7 Demanding customers

The industrial revolution not only produced many new construction technologies, it produced ever more demanding customers. Many of these owned or managed large commercial companies and expected new facilities to be produced quickly and reliably. Governments also often needed new facilities quickly to meet changing social needs, implement new policies or fight a war. These demands became the norm and individuals followed the lead of companies and government agencies in expecting construction to be efficient and reliable. The new demands influenced perceptions of time, cost and quality. They also changed the nature of the new facilities customers expected construction to provide.

As far as time was concerned, customers wanted their new facilities as quickly as possible. They certainly wanted them constructed faster than fundamental traditional construction normally achieved. They also wanted to be sure the agreed completion date would be achieved so they could make firm plans for moving into and using the new facility.

Customers also wanted lower costs. Commercial companies in particular expected to negotiate low prices for the resources they needed and wanted their construction projects to be undertaken efficiently so costs were kept to a minimum. They also wanted to know exactly when payments would become due and the amounts which would need to be paid so they could make the necessary financial arrangements. These new demands were led by commercial companies but government bodies and individual customers rapidly followed suit because they all needed to plan their finances.

Demands for reliable quality resulted from customers expecting new facilities to be complete and working exactly as they had agreed. They wanted this reliable quality from the day they took possession of their new facility. They also wanted assurances that any defects which became apparent in the first months or years following completion would be put right without hassle or additional costs.

Beyond all this customers wanted their new facilities to provide ever more comfort and convenience. This might be to provide more comfortable places to live or better working conditions, facilitate the movement of people and materials, or any other innovation which contributed to the effective use of the new facility. Customers also wanted their new facilities to be safe for everyone using them. Beyond this many customers expected new facilities to provide a symbol of their own worth and importance.

The new demands reflected a growing confidence fed by rapid advances in science and engineering. Humans began to feel they were masters of their environment and expected construction to reflect this. No doubt customers had always wanted construction to be efficient and reliable. The difference in the approach of the more demanding customers produced by the industrial revolution was they insisted on their demands being incorporated in contracts. This ensured that the companies producing their new facilities would have no doubts about what they were required to deliver and if they failed could expect the customer to demand financial compensation.

In terms of the theory of construction management, the required level of efficiency was increased by more demanding customers. This inevitably raised the inherent difficulty faced by construction project organizations based on the methods and standards of fundamental traditional construction.

5.8 Developed traditional construction

The responses to new technologies and demanding customers provide the essential features of the developed form of traditional construction which is implicitly or explicitly assumed in most modern construction management literature. It is embodied in the professional roles and commercial relationships assumed by most of the governing bodies and institutes which control national construction industries. Specific features of the developed traditional approach vary from country to country but the essential features are broadly similar because they have developed from very similar fundamental forms.

The developed form of traditional construction begins with a construction customer employing a design consultant. The first essential activity in traditional construction is to prepare a brief which describes what the customer wants in the new facility. The designer guides the customer into considering all the practical issues raised by the construction of the new facility and ensures the likely timings and costs are carefully evaluated. Traditional construction works most effectively when the customer is clear and certain about their new facility and its implications for their organization and ensures this is all described in the brief.

When this process is carried out effectively, the brief provides a formal description of the new facility in the customer's own terms. This is supported by a clear statement of the realistic time and costs the customer can afford to allocate to the production of the new facility.

As the brief is developed, the primary design consultant begins design. This is a complex problem-solving process requiring many, often conflicting requirements and constraints to be reconciled. These include the customer's brief and the designer's ambitions to produce a new facility which looks good and functions well. At the same time they need to take account of the peculiarities of the site selected for the new facility, competing demands from powerful interests both inside and outside the customer's organization, and many physical and technical issues as well as the time and cost constraints imposed by the customer.

The number and complexity of the issues to be considered means design is necessarily a sequential process. It begins with broad concepts, develops into an overall scheme design which is expanded by engineering the individual systems which enable the facility to meet all the customer's needs, and finally moves into the detail design stage.

The design of most new constructed facilities involves a variety of design specialists. Building projects, for example, normally need structural and services engineers and others providing specialised design knowledge and skills. Engineering projects need different types of specialised design knowledge and skills but are often every bit as complex as major building projects.

The design specialists are employed on the advice of the primary design consultant. Provided they are allowed to work within their normal professional boundaries, traditional design proceeds steadily and reasonably efficiently. The results of each stage are recorded in ever more detailed drawings and specifications which describe all the parts of the proposed new facility.

The next stage in the developed form of traditional construction is to select and employ a general contractor. This important development resulted from customers, frustrated by the uncertainty of fundamental traditional construction, negotiating contracts with one construction company to take responsibility for ensuring the new facility was completed by an agreed completion date for an agreed sum of money.

Initially, many of the companies selected to play the role of general contractor were those responsible for the technology which provided the main structure of the new facility, usually masonry, brickwork or timber. Over time some of the companies given this leading role abandoned their individual specialist activities and concentrated on acting as general contractors. Others continued to use directly employed workers to undertake significant production activities. The exact pattern of subcontracting and direct employment varies widely as it is influenced by national conventions and legislation, and local labour market conditions.

Whatever pattern of direct employment and subcontracting they adopted, general contractors distinctive skill was in organizing the procurement, manufacture, production and commissioning of all the resources needed to produce new facilities. They learnt how to coordinate these activities so they could agree firm contracts with customers to deliver a new facility in accordance with a given design on time and cost. As they became more experienced in coordinating and more powerful in dictating terms to manufacturers and assemblers, general contractors found ways of completing construction projects more quickly and efficiently than had been the norm with the old separate trades approach.

The selection of a general contractor for an individual project is normally organized by the primary design consultant on behalf of the customer. It often involves a competitive bidding process in which a number of general contractors offer to complete the new facility by the required date for a stated sum of money. The bids are evaluated by the primary design consultant who recommends the employment of one of the general contractors. Within most construction industries, this is normally the one submitting the lowest bid.

Once appointed the general contractor analyses the design information and plans the manufacturing, production and commissioning activities. These are then put into effect by the general contractor selecting and employing companies to undertake any specialised manufacturing, supply general materials and undertake the production and commissioning activities.

The manufacturing, production and commissioning activities are supervised on behalf of the customer by the design consultants. The aim of this supervision is to ensure the quality of the new facility but not in any way to assume responsibility for the general contractor's performance. This supervision continues until the primary design consultant certifies the new facility is complete.

5.9 Internal and boundary relationships

The developed approach to traditional construction is put into practice by a project organization which emerges in two distinct stages. This is illustrated in Figure 5.2.

The project organization begins with the formal boundary relationship between the customer and design consultant. The initial stage develops as specialist consultants are employed to advise and assist the primary design consultant. Some of the consultants may have established relationships with the

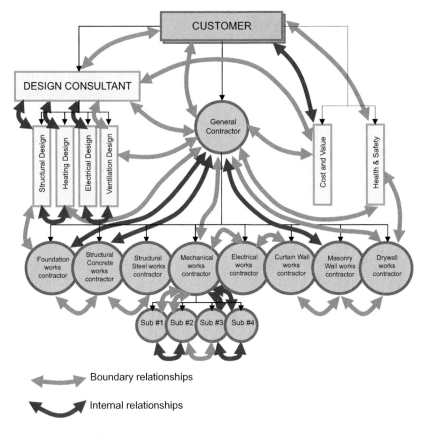

Boundary relationships

Internal relationships

Figure 5.2 Developed Traditional Construction Project Organisation.

primary design consultant but others, some selected by the customer, result in more boundary relationships.

Parts of the initial design-consultant-led organization persist throughout the project. However, once the general contractor is employed, the effective project organization changes. Manufacturing and production activities require the general contractor to employ specialist companies leading to the formation of a separate project organization organized by the general contractor.

The general-contractor-led project organization works through a mixture of boundary, established and internal relationships. The actual pattern is partly dependent on the general contractor's experience of the specific type of new facility being produced and the extent to which they use directly employed construction teams. It is perhaps more dependent on the number of innovative design details and new technologies included in the design. It is also influenced by the local construction market and the extent to which the main contractor is able to agree contracts with their preferred specialist companies.

The design-consultant-led part of the overall project organization adopts a supervisory role over the work produced by the general-contractor-led part. Inevitably the interactions between the two parts of the project organizations take place through formal boundary relationships almost entirely shaped by the terms of the individual contracts which brought them into the project organization.

5.10 Strengths of developed traditional construction

Developed traditional construction works well for everyone involved when the customer has a clear view of the new facility they want, employs a designer who concentrates on producing a design guided by the customer's demands which uses technologies well within the competence of local general contractors and the manufacturing, production and commissioning companies they employ.

In this ideal approach, each team knows what to expect from those whose work precedes their own and they know how to leave their own work so as not to create problems for those whose work follows on. This enables design consultants to work with predetermined plans of work and production teams to work with predetermined sequences of work. Teams can, to a large extent, work independently of each other. It is practical to define the work of each company involved in sufficient detail to give a secure basis for standard contracts to be used successfully.

When developed traditional construction is well established, local construction markets are likely to include many firms competent to undertake each type of work. It is therefore sensible to select firms by means of straightforward competitive bids. This may bring together teams who have not worked together before but their work will still be coordinated by established custom and practice supported by predetermined plans and sequences of work and standard forms of contract which reflect established ways of working.

When these conditions are met, the results can be outstandingly good. Traditional construction has produced many of the most widely admired buildings and engineered infrastructure. Designers have achieved world class reputations by respecting traditional construction's limitations. Many construction companies have created successful businesses based on competence at traditional construction. Customers get a fair deal and in general are satisfied with their new facilities.

5.11 Weaknesses of developed traditional construction

The outcomes for developed traditional construction are very different when any of the conditions for success break down. The results tend to be unsatisfactory, even disastrous, for at least some of those involved. Work is delayed, costs escalate and quality is compromised.

There are many possible causes of unsatisfactory results. Customers may have unrealistic aims for their new facility. They may change their mind as they see it being assembled or learn the time and costs involved. Designers may produce designs which are beyond the competence of local manufacturing and construction companies. Designs may be incomplete at the time the general contractor is appointed. Design consultants may change the design during the manufacturing and production stages resulting in abortive work and increased costs. The outcome is the vicious circle illustrated in Figure 5.3.

These design failures often result from misunderstandings and problems within the design teams assembled around the primary design consultant. This

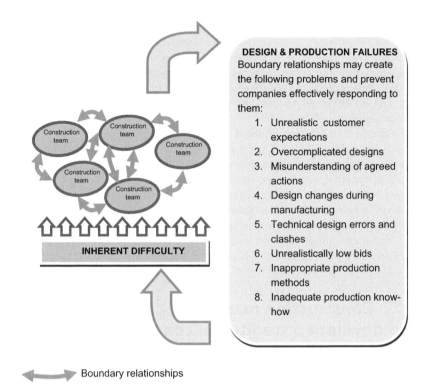

DESIGN & PRODUCTION FAILURES
Boundary relationships may create the following problems and prevent companies effectively responding to them:

1. Unrealistic customer expectations
2. Overcomplicated designs
3. Misunderstanding of agreed actions
4. Design changes during manufacturing
5. Technical design errors and clashes
6. Unrealistically low bids
7. Inappropriate production methods
8. Inadequate production know-how

Boundary relationships

Figure 5.3 A Vicious Circle of Boundary Relationships in Developed Traditional Construction.

is particularly likely on building projects because of the distinct cultures developed by architects and engineers. On all except the simplest of buildings, the architect as primary design consultant is likely to need structural and services engineers, specialists in external cladding systems, interior design and many more. These specialist engineers are educated and trained in ways which are based on projects where engineers are the primary design consultants. These are projects to construct the physical or services infrastructure or new industrial facilities.

The important effect of these different histories is the two primary design professions have developed independently with architects having a primary focus on the appearance of buildings and engineers having a primary focus on the technical competence of new facilities. This fundamental difference in viewpoint means the two professions relate to customers in distinct ways. They make different demands on other construction companies. All this is reflected in efforts to use specific forms of contract designed to establish their preferred ways of working. Unfortunately some of the most widely used standard forms of contract include terms and conditions which conflict with terms in other standard forms. Where these are used on the same one project, the problems are inevitable and can all too easily be magnified instead of being solved quickly and amicably. The clash of distinct professional cultures and ways of working has the potential to create misunderstandings and problems throughout all stages of projects not just during the design.

The project situation is further complicated once the general contractor is selected. They may have submitted an unrealistically low bid to win the contract. They may have misjudged the nature of the manufacturing and production activities required by the design. They may experience difficulties meeting design consultants' demands. In an effort to rescue their own position, they will be tempted to pass at least some of these problems to their subcontractors.

Whichever combination of these possible problems affects an individual project, manufacturing and production companies often find themselves locked into contracts which leave them to choose between using the cheapest possible materials and components and poor quality standards, or losing money and completing the work late. Inevitably the customer loses out. They are likely to be provided with a facility which does not meet their needs, works badly and includes at least some elements of poor quality. More than this it is likely to be completed late and involve the frustrated customer in additional costs and very possibly formal disputes and litigation.

5.12 Construction management propositions and developed traditional construction

Developed traditional construction essentially represents a strategy of simply accepting the inherent difficulty. This cannot sensibly be called a construction management strategy as it effectively ignores the construction management propositions.

For the majority of modern construction projects this is not an effective approach. As described earlier, new technologies have increased the complexity faced by project organizations and demanding customers expect construction to deliver increased levels of efficiency. These new challenges cause difficulties for developed traditional construction which frequently cause it to fail. Success depends on customers and their designers requiring well-established technologies to be used in conventional ways. This can give the general contractor role a realistic chance of providing sufficient coordination of the procurement, manufacturing, production and commissioning activities to allow all the companies involved to undertake their activities efficiently.

The chances of these circumstances occurring are increased by companies specialising in a narrow range of projects and working with a small number of customers and other construction companies. In many construction markets, this narrow specialisation is difficult to sustain and it is more common for developed traditional construction to fail. The fundamental problem is the pattern of formal boundary relationships formed by developed traditional construction does not work at all well on complex or uncertain projects. They are put under considerable strain which results in reduced efficiency. This may be caused by ineffective coordination by the primary design consultant or during later stages by the general contractor. Alternative problems often result directly from boundary relationships having insufficient capacity to deal with difficult problems. In any of these situations, the result tends to be misunderstandings and confusion.

A common response is the emergence of de-facto teams working through internal relationships as individuals find it quicker and more efficient to communicate directly with others working on related aspects of the new facility. However, such informal internal relationships are rarely fully effective. Often this is because the internal relationships are not entirely trusted especially when they conflict with contract provisions which require interactions to take place using specific formal routes. For whatever reason, informal relationships tend to break down if problems are not resolved quickly. When this happens, construction teams revert to the security of formal boundary relationships. Indeed faced with serious problems, consultants or construction companies may bypass the central role of the primary design consultant or general contractor and insist on dealing directly with the customer.

The root cause of many of the failures of developed traditional construction lies in the mismatch between design consultants producing a complete design and much of the knowledge and skills needed to design practical systems and details being owned by specialist construction companies. Developed traditional construction relies on the assumption that the general contractor is selected and employed on the basis of a complete design. This assumption is unrealistic since the specialist construction companies competent to design practical systems and details are employed and brought into the project organization by the general contractor.

Overall developed traditional construction establishes a project organization dominated by formal boundary relationships, constrained by contractual terms and conditions and guided more by a need to protect individual interests than by any concern for the customer's interests. Informal internal relationships develop as problems arise because those involved hope it will help them find a solution which they can accept as good enough. But these tend to break down if quick solutions to problems cannot be found. It has to be concluded that on all except straightforward and certain projects, developed traditional construction provides an inadequate management system. This is reflected in the example project which follows. The example also serves to illustrate the use of the Inherent Difficulty Indicators introduced in Chapter 4.

Example – Building Project using Developed Traditional Construction

The project used in the example is a new hotel on a virgin site. The hotel has 300 elegant, well fitted and equipped guest rooms and a wide range of other spaces to provide everything expected in a 5-star hotel. The building is 10 stories high and has a steel structural frame which supports concrete beam floors and roofs, and high quality, prefabricated external cladding. It has sophisticated service installations including heating, ventilation and air conditioning, electrical and communication networks, security systems, hot and cold water supply networks, robust drainage systems, and so on. Internal divisions are provided by a robust, prefabricated partition system. It has an impressive entrance and sophisticated elevators. The internal decoration is elaborate and stylish. The hotel stands in extensive grounds which include extensive car parking, tennis courts and attractive gardens.

Basic data describing the project

	Number of teams	Number of team-days
Brief	1	80
Design		
Concept design	1	100
Scheme design	1	80
Detail design	6	100
Technical design	6	150
Production information	2	120
Designer's supervision	2	180
Plan	3	200
Procurement		
Select companies	7	300
General contractor plan and control	4	350
Contracts and payment systems	3	100
Manufacturing	5	150
Production		
Substructure	4	100
Structure	7	210
External envelope	6	300
Service cores	6	150
Risers and main plant	5	120
Entrance and vertical circulation	12	500
Internal divisions	3	120
Decoration	12	750
Fittings	6	450
Landscaping and external services	6	250
Commissioning	7	140
TOTALS	115	5,000

Calculation of the inherent difficulty indicators

From equation 4.11 there are $R = 6,555$ possible relationships between teams working on this project. The detailed project plan, which can be seen on the website linked to this book, www.wiley.com/go/construction-managementstrategies, shows that 3,646 of the possible relationships do not occur because there is no overlap between activities. This gives a total of $R-k = 2,909$ direct interacting relationships that do in fact occur.

The largest number of teams which could be involved in the same time interval is 47 in interval 21 so there could be 1,081 different relationships within that time interval. However, since not all teams work at the same time it is imperative to determine the number of teams which work simultaneously for each time interval. To relate it back to this example, the plan shows that during the 21st time interval the following teams work simultaneously:

- 3 out of 6 external envelope teams;
- 5 out of 6 service core teams;
- 5 out of 5 risers and main plant teams;
- 4 out of 12 entrance and vertical circulation teams;
- 2 out of 6 landscaping and external services teams;

- 2 out of 3 contracts and payment system teams; and
- 2 out of 4 general contractor's plan and control teams.

This gives a total of 23 teams with 253 relationships.

The above exercise shows just how important it is to determine, early in the project, the number of teams which will work simultaneously. Examining similar time intervals in past projects is a very good starting point but construction managers should not rely solely on past data and should take into account specifics of the project under consideration which may deviate from those in the past to obtain as accurate picture as possible.

There are two 'Structure' teams which have established relationships with three 'External Envelope' teams, and six 'Service Core' teams which have established relationships with five 'Risers and main plant' teams. This gives a total of 65 established relationships between 15 teams and an Established Relationships Indicator: $E_R = 65/2,909 = 0.022$ (equation 4.13). This is a very low value which is then further confirmed by equation 4.14 which gives time dependent indicators of established relationships ranging from 0 to 0.05. Established relationships indicator for an individual time interval should include all possible relationships and not just those between teams which are supposed to work simultaneously because established relationships may exist even between teams which are not planned to work simultaneously.

The relationship fluctuation indicator F_E is calculated using equation 4.15 and gives a value of 0.985. We can see a value of F_E close to 1 and a very low value of E_R, which reflects the overwhelming influence of boundary relationships throughout the project because there are few established relationships. As a consequence standard deviation is very low because most values of E_{Rv} are zero so F_E and E_R should always be examined in tandem.

However, to really establish the benefits of established relationships we have to look at how long the teams involved are working together on this project and how much time they have spent working together on past projects. Then equation 4.16 gives a Relationship Quality Indicator for each established relationship which ranges from 0.01 to 0.85 and the total Relationship Quality Indicator (equation 4.17) is: $Q_R = 28.85/2,909 = 0.0099$. Although there are a few high values for individual Relationship Quality Indicators, the total Relationship Quality Indicator is very low because there are only 65 established relationships in this project out of 2,909 which actually occur.

According to equation 4.18 we can calculate the relationships configuration complexity indicator C_R (note that $\alpha = \beta = \gamma = 0.333$):

$$C_R = \tfrac{1}{3}\left(\left(1 - \tfrac{29}{229}\right) + \left(1 - \tfrac{77.3}{2120}\right) + \left(1 - \tfrac{23}{115}\right)\right) = 0.88$$

The indicator shows the project organization is not overly complex. There are 29 time intervals but there could be a maximum of 229 time intervals. On the other hand, this is balanced in terms of team congestion as the maximum number of teams working at any given time is 23 in the 21st time interval. Time intervals are also relatively short in comparison to total project duration, which in the case of boundary relationships may play a crucial role in preventing escalation of possible problems which may occur if time intervals are much longer.

The past performance of the 115 teams over their immediately previous 10 projects provides individual team Performance Variability Indicators which range from 0.10 to 0.50. These give a total Performance Variability Indicator (Eq. 4.20): $R_p = 0.24$.

This is of course the average performance variability showing the teams managed to complete their work within the agreed time on only 24 per cent of their recent projects (note we have assumed c_j to be 1 for all teams although for real life projects these values should be determined separately).

The External Interference Indicator based on historical records for the particular region for the same type of building project is 0.45.

Overall, we have the following inherent difficulty indicators:

Established Relationships: $E_R = 0.022$ ($E_{R,max} = 1.0$)

Relationship Fluctuation: $F_E = 0.985$ ($F_{E,max} = 1.0$)

Relationship Quality: $Q_R = 0.0099$ ($Q_{R,max} = 1.0$)

Relationship Configuration: $C_R = 0.88$ ($C_{R,max} = 1.0$)

Performance Variability: $R_p = 0.24$ ($R_{p,max} = 1.0$)

External Interference: $I = 0.45$ ($I_{max} = 1.0$)

The low value of the indicators for Established Relationships, Relationship Quality, Performance Variability and External Interference suggests the project is inherently difficult. High values of the indicators for Relationship Fluctuation and Configuration should be approached very cautiously. The very low number of established relationships deems the project inherently difficult regardless of the highly optimised configuration and low relationship fluctuation as these cannot offset the fact that most teams on this project have never worked together before. There is therefore a high probability the project will not be completed as planned.

This reflects the teams' relatively poor performance record and although there are 15 teams with established relationships, they benefit only a small proportion of the interactions. Given the likelihood of external interference, it is clear that the project organization's decision to use developed traditional construction gives them a difficult task in attempting to complete the project as planned.

5.13 Scenarios to rescue developed traditional construction

The example highlights the weaknesses in developed traditional construction. This chapter concludes by describing three significant approaches which seek to deal with the major failures of developed traditional construction while preserving its essential characteristics. The three approaches serve to highlight the nature of key characteristics of developed traditional construction.

The first development of developed traditional construction goes to remarkable lengths to maintain the architect's powerful, central role in building projects. The approach devised by architects in the United Kingdom provides the clearest example of this. The second scenario limits the architect's role to producing a scheme design and accepts that the technical knowledge and skills needed for detail design is owned by specialist manufacturing and production companies. The developed traditional approach in the United States illustrates this approach. The third scenario is similar to the United Kingdom approach in seeking to maintain the architect's powerful, central role in building projects. However, it is similar to the United States approach in fully recognising that the technical knowledge and skills needed for detail design is owned by specialist manufacturing and production companies. The developed traditional approach found in many European countries illustrates this approach.

5.14 Projects led by design consultants

Architects quite properly want to produce beautiful buildings. In the United Kingdom many of the most influential architects believe the architectural integrity of their designs depends on them having control of all the design details. The United Kingdom's traditional building industry has devised organizational arrangements to enable architects to exercise this control. It creates two distinct stages which are reflected in the project organization as illustrated in Figure 5.4.

The central innovation of the approach is to enable architects to work with specialist construction companies during the design stage before the general contractor is selected. These are specialist construction companies with knowledge and skills of the technologies which will be used in the building and so they are able to guide the architect in producing practical design details.

The specialist construction companies are rewarded for their work either by being paid to act as a design consultant or the more usual approach is for the architect to nominate or name them in the general contractor's contract. The effect of this is they are employed as a subcontractor once the general

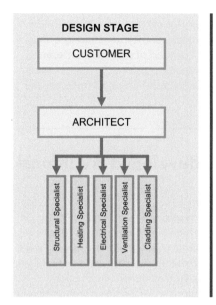 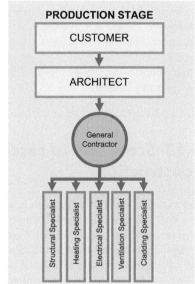

Figure 5.4 Main Stages in the United Kingdom Developed Traditional Construction.

contractor has been selected and employed. This approach requires the specialist construction companies to undertake design on trust in the expectation they will subsequently be employed to carry out the resulting manufacturing and production work.

Both ways of rewarding the specialist construction companies can create problems. Many specialist construction companies are not well organized to act as consultants. Their knowledge and skills are centred on manufacturing and production activities. They undertake design in support of their manufacturing and production activities and this does not equip them to participate in a multi-discipline design team. Also it is not uncommon for the fees paid to specialist construction companies to be derisory, based on the argument that they are in a strong position to win the contract for the manufacturing and production work. Nominated or named subcontractors are in an even weaker position since they have to hope their design work will be properly reflected in the subcontract price. Their situation can be worse than this. They may be required to enter into a contract with a general contractor they do not know. Even worse, it can happen that the selected general contractor is one they are in dispute with on other projects. Almost all the possible ways of resolving such unfortunate situations are unsatisfactory.

A related weakness of the United Kingdom approach is it tends to create ambiguous relationships particularly for specialist construction companies. At one point in the project they are working under the direction of the design consultant but subsequently they become a subcontractor to the general contractor who is being supervised by the design consultant. When design issues arise during manufacturing or production, it is difficult for the specialist construction companies to know who they should deal with. Unfortunately, in practice it is often the case that at least some of those involved decide it is in their interests to ferment the confusions and doubts in the hope of establishing a claim for additional money or time.

The United Kingdom approach can also cause problems for manufacturing and production companies who are not involved during the design stages. They may find themselves faced with design details, produced by the architect in isolation, which are difficult or even impossible to manufacture or assemble within the times and costs provided for in their subcontract. Indeed they may find they have bid for the work on the basis of designs which are radically changed by the architect after work has commenced and then discover they are locked into contracts which give them little scope for recovering the full costs of the consequent disruptions.

The overall result is the arrangements set up to give the architect control over the detail design tend to become confusing and inefficient in practice. This has given rise to a uniquely British profession, that of quantity surveyor. Their essential task is to produce bills of quantities which list the materials and components, and the production and commissioning work required to produce the design which forms the basis of the contract between the customer and general contractor. Bills of quantities are produced on the basis of rules of measurement agreed by organizations representing general contractors and quantity surveyors.

Bills of quantities allow competitive bids to be submitted on an exactly comparable basis thus facilitating the selection of a general contractor. Once the general contractor is selected, bills of quantities provide the basis for the contract with the customer. Then bills of quantities provide an administratively straightforward means of calculating the price for any changes the architect may decide to make during the manufacturing, production and commissioning stages of the project.

These arrangements are often abused especially when an architect is struggling to complete the design to meet the customer's timetable. A common device is to base the bills of quantities on vague or incomplete design information. This allows competitive bids to be obtained but means the general contractor is misled about the nature of the work they will be required to organize. The most likely outcome is a project beset by disputes and claims.

The main benefits provided by the quantity surveyors role occur during the design stages. Working with bills of quantities priced in competition gives quantity surveyors an extensive data base of local construction costs. This has been developed into cost planning processes which can be effective in guiding the design to match the customer's budget.

Cost control and any other controls over design depend on the customer being resolute in insisting the control systems are respected by designers. Without this support, the powerful position of the primary design consultant is too often used to emasculate any control systems.

This is but one example of the way the architect's role within the United Kingdom developed traditional construction is negotiated from project to project. Experienced architects begin establishing their power base during their first discussions with the customer. Ambitious designers will bombard the customer with drawings and models of beautiful, exciting, leading edge designs. They may encourage the customer to visit outstanding examples of the type of facility they want. The overall aim is to enlarge the customer's ambitions sufficiently to allow the design consultant to concentrate on producing a highly creative design. This suits designers' culture but many customers have difficulty understanding designers' exciting ideas and struggle to ensure their own needs are taken into account in the brief.

A weak brief is likely to lead to failures in the design which are not apparent to the customer until the new facility begins to emerge on site. This may suit a designer determined to maintain their freedom to produce designs which satisfy their own aspirations. Even if the customer employs a quantity surveyor to exercise cost control, a determined designer can undermine the process by prevaricating about design decisions. This is a recipe for problems in subsequent stages as the customer learns the costs of their exciting but inadequate new facility.

Overall the United Kingdom developed traditional construction produces some outstanding architecture which stands out from the variable quality evident in the majority of the United Kingdom buildings. However, the approach is also characterised by inefficient manufacturing, production and commissioning, late completions and high costs caused in no small part by the unsatisfactory situations created for manufacturers and producers. Some are employed on the basis of ambiguous and vague contracts, while others are forced to wrestle with impractical design details. This represents a high cost for the occasional piece of outstanding architecture.

In terms of the theory of construction management these outcomes result from:

- an increase in the number of construction teams due to the introduction of quantity surveyors; and
- an acceptance of the agreed objectives of the project by the construction teams involved and the quality of communications between them being influenced by the extent to which the bills of quantities provide an accurate description of the required work.

5.15 Specialist contractor design

The traditional building industry in the United States is in essence fundamental traditional construction plus general contractors. Architects accept that much of the detail design is best produced by specialist manufacturers and assemblers who understand the technologies involved. Architects therefore concentrate on producing a scheme design which meets the customer's needs. They have devised sophisticated systems capable of producing drawings and specifications quickly which describe their scheme designs sufficiently to enable general contractors to submit competitive bids. Having reached this stage in a project, architects do not alter the scheme design.

Once selected, the general contractor agrees a contract with the customer to produce a building in accordance with the scheme design. This is achieved by employing subcontractors competent in all the technologies required to produce the building. The subcontractors select detail designs consistent with the scheme design. They can do this quickly and efficiently because they have libraries of design details to suit the kind of buildings designed by local architects. They understand how their design details relate to those used by other subcontractors and are competent in undertaking the manufacturing, production and commissioning activities required by their own design details.

The architect is often retained by the customer to check the detail designs proposed by the subcontractors. This is to ensure they reflect the scheme design reasonably well but not to change or develop it. It is the general

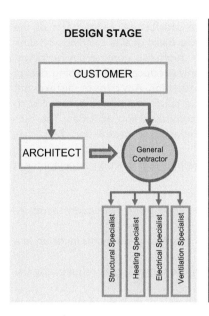
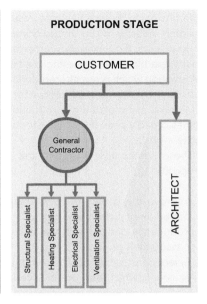

Figure 5.5 Main Stages in the United States Developed Traditional Construction.

contractor who is responsible for producing the building in accordance with the scheme design. This includes any design work needed to produce a complete building.

As a result of subcontractors working on designs which suit their own knowledge and skills, manufacturing, construction and commissioning normally proceed very quickly and smoothly. Problems on construction sites are sorted out immediately by the construction teams directly involved. The overall emphasis on speed and economy means some design details fit together in an awkward, even an ugly manner. In general this is accepted provided projects are completed on time and within budget and the completed building functions reasonably well. The subtle changes in responsibilities within the project organizations established for each of the two main stages are illustrated in Figure 5.5.

The United States approach hits problems when designs make unusual demands on local manufacturing, production and commissioning companies. Manufacturing can be delayed as unexpected problems have to be resolved often requiring reference back to the general contractor, designer and even the customer. Similarly, work on site can become disorganized and inefficient. Quality tends to suffer. Where the details, in the completed facility are hidden by cladding or finishings, untidy workmanship is accepted and defects are patched up. It is also the case that commercial pressures often mean awkward, even ugly, details are accepted despite them being visible in the completed facility. Similar compromises are accepted in respect of time and cost as tough commercial pressures drive everyone involved to maintain a fast rate of production. Almost inevitably, many individually designed projects end up as disastrous failures for everyone.

The overall picture is that the developed United States approach to traditional construction provides variable results. Where the scheme design is within the established competence of local manufacturing, production and commissioning companies, the results are satisfactory. Large parts of the building industry in the United States deliver basic quality buildings on time at low costs. For example, supermarkets, high rise commercial buildings, apartment blocks and housing are commonly produced reliably using design information which is not altered during construction. Any gaps or mistakes are put right by the general contractor or the subcontractors they employ to undertake manufacturing, production or commissioning activities. The outcomes broadly provide what customers want and the construction companies make a profit.

These outcomes support the following construction management propositions:

- Improve the quality of relationships between the construction teams involved in a construction project.
- Reduce the performance variability of the construction teams involved in a construction project.
- Select construction teams competent in the technologies required by the project in which they are involved.
- Ensure construction teams accept the agreed objectives of the project in which they are involved.
- Ensure construction teams are motivated to achieve the agreed objectives of the project in which they are involved.
- Foster accurate communication between the construction teams involved in a project organization.

However, where the scheme design is not within the competence of local manufacturing, production and commissioning companies, the outcomes result from a neglect of the construction management propositions. Such projects frequently end up in court as everyone involved tries to rescue their own financial situation from a disastrous project.

5.16 Architects and engineers design

Much of Europe has retained an essentially traditional approach in which customers appoint an architect to produce the brief and scheme design. When a design has been agreed, the customer appoints an engineering consultant or contractor to select specialist contractors who undertake the detail design based on their preferred technologies and ways of working. Many of the specialist contractors have an engineering background which helps them produce technically competent detail design proposals. When the customer agrees a detail design, the specialist contractors organize any manufacturing, and undertake the production and commissioning. The two main stages of this approach are shown in Figure 5.6. This approach can result in a successful project when the specialist contractors work to the architect's original design.

In terms of the theory of construction management these outcomes result from ensuring that construction teams are competent in the technologies required by the project. This tends to ensure the construction teams accept the agreed objectives of the project.

In practice the specialist contractors often make considerable changes to the architect's original design in order to make best use of their preferred

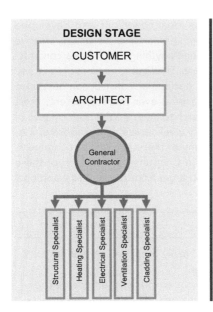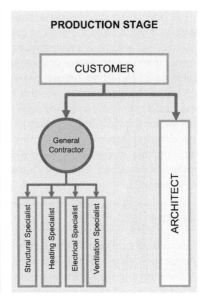

Figure 5.6 Main Stages in the European Developed Traditional Construction.

technologies and ways of working. There are many possible causes for this potentially dangerous practice. The architect's design may go beyond the capability of established local technologies and specialist contractors. It may also result from an inappropriate choice of specialist contractors. Often the fundamental problem is the specialist contractor's engineering background leads them to misunderstand the architect's scheme design. Whatever the causes, the results often include disputes with customers who are entirely satisfied with the architect's scheme design and expect the specialist contractors to put it into practice. Worse outcomes include construction failures as detailed designs produced by individual specialist contractors turn out to be incompatible. This frequently results from the specialist contractors interpreting the scheme design in different ways. In any of these unfortunate situations it is common for the customer to re-employ the architect on a new contract to bring the project back on track. This practice frequently leads to battles between architects and engineers because it is often difficult to prove who caused the problems. These outcomes result from a failure to act on the basis of the construction management propositions. In particular construction teams are selected that are not competent in the technologies required by the project and they have not fully accepted the agreed objectives of the project.

5.17 Conclusions

Developed traditional construction is widely used and when customers and construction companies use well-established technologies in conventional ways, and the general contractor is effective in coordinating the procurement, manufacturing, production and commissioning activities, all the companies involved can undertake their activities efficiently.

It is difficult to create and sustain these ideal conditions and developed traditional construction is weak at coping with demanding customers, creative designs, technological innovation, and indeed anything which faces construction companies with new or unusual situations. The limited success of the three scenarios intended to rescue developed traditional construction serves to emphasise its weaknesses. As this has become ever more apparent, most developed countries have undertaken research to measure the performance of their national construction industry, identify weaknesses and propose strategies aimed at improving performance. Many of these proposals recommend adopting one of the scenarios described in Chapters 6 to 9. Each of these modern approaches represents a distinct change from developed traditional construction.

Exercise

Consider the project to produce an individually produced traditional brick house for a single family described below.

Building the house involves a number of specialist craftsmen experienced in undertaking construction actions on individual house projects. Most activities on the house building project will be undertaken sequentially with just a few being performed in parallel so in general there will only be limited interactions between trades for relatively short periods. The activities are interconnected in a logical construction sequence so you will have to consider the following features of established local construction:

1. Architectural drawings (this would produce a complete design which gives the production teams sufficient information to undertake their work. The architect is often employed to check the construction to make sure the design is being followed).
2. Management (the individual housing sector is dominated by customers directly managing the project but some employ an external project manager who manages day-to-day activities on site).
3. Construction site preparation (this include clearing the vegetation and topsoil, grading and securing the site perimeter).
4. Foundations (this may include several trades including those responsible for formwork and concrete but more often this will be produced by a single team of experienced foundation workers).
5. Brick wall construction (external and any internal brick or block walls are constructed by a team of skilful bricklayers and their work normally includes installing external or cavity-wall insulation).
6. Roof (teams of carpenters, roofers and tilers are employed to erect, insulate and tile the roof).
7. Electrical works (a single family house does not require an extensive electrical installation so a single team of skilful electricians can undertake the complete job).
8. Plumbing (this would require plumbing installations in a kitchen and bathrooms but depending on the type of heating installation a certain amount of plumbing work may be required for the heating installation).

9. Drywall installation and finishing (internal walls may be of drywall timber-frame or block construction lined with drywall boards undertaken by a team of specialist installers and finishers).
10. HVAC installation (a modern family home may have a relatively complicated HVAC system with integrated heating, ventilation and air conditioning but a single team of specialist HVAC installers would be able to complete the job in conjunction with plumbers).
11. Painting and decorating (once all the services are installed and walls completed, a single team of skilful painters and decorators could paint the internal space and install decorative elements).
12. Landscaping (a family house would not be complete without carefully designed surroundings involving specialist landscape and garden designers).

The relatively limited number of teams and interactions involved in building a family house suggests a traditional or developed traditional construction approach would be adequate. There are rarely more than two or three teams working on the construction site simultaneously, so any misunderstandings and errors can easily be resolved on the spot. The customer would also be able to explain requirements in detail so the architect could produce a complete design prior to start of construction. Any omissions or errors would be easily resolved with the help of skilful specialist craftsmen even if the architect is not present during construction.

How does that example compare to a 300-room hotel used as the basis of the Example project described earlier in the chapter? Would a traditional construction approach in any form be sufficient to deal with complexities of all the multiple team interactions on the hotel project?

Consider the construction implications of all this and think about the following questions:

1. Look at the above information given in the Example and calculate how many different construction teams would be required to build the hotel. In making this calculation remember there are 300 guest rooms and many other spaces in the hotel so think how many teams for each trade would you need to complete the work if only two different trade teams can work in one guest room at the same time and at the most 100 rooms would be accessible for work at any one time.
2. Contrast the requirements an architect would need to consider to produce a detailed design for the hotel with that required for the single family house. Would a customer for a new hotel be able to provide the same level of detail as a customer for a family house (think about the different kinds of rooms needed, the comfort and access demands of guests, the ventilation, temperature, lighting and other regulations that determine requirements for a 5-star hotel)?
3. It is likely that dozens of teams would be working on site at the same time. How would you manage such a project in order to ensure a smooth sequence of activities?
4. What would you need to do if several different manufacturing and production teams report design errors?

5. Would it be possible to successfully deliver the hotel project through the developed traditional construction on time and within budget?
6. Consider the three developed traditional construction scenarios in the main body of text in this chapter and decide if any of them provide an approach more likely to ensure successful delivery?

Further Reading

The following publications are the source of ideas used in this chapter and provide further information for readers.

Hillebrandt, P. M. (1984) *Analysis of the British Construction Industry.* Macmillan. This book provides the outstanding description of the United Kingdom construction industry at a time when it was still dominated by developed traditional construction. The detailed accounts of past output data, industry structure and prevailing processes enable comparison with current developments in the industry, which is invaluable for understanding how construction has evolved over the past few decades.

Sabbagh, K. (1989) *Skyscraper: The Making of a Building.* Macmillan. This popular book was supported by a fascinating series of television programmes. It provides an engaging description of the United States approach. The author uses an engaging and playful writing style to uncover the realities behind the making of a skyscraper from the standpoints of involved individuals and companies.

The Tavistock Institute (1966) *Interdependence and Uncertainty: A Study of the Building Industry.* Tavistock Publications. This report describes rigorous social science and operational research into traditional construction. It explains how traditional construction operates and how the main roles interact. By looking at operational and socio-technical systems this report highlights perennial problems by using easily accessible language in a concise and well-structured way. The last part of the report focuses on further research and highlights topical issues from which researchers could draw further ideas.

Chapter Six
Design Build

6.1 Introduction

Design build provides an attractive approach for construction customers frustrated by the failures of traditional construction. Design build enables a customer to enter into a contract with a design build company to produce a new facility for an agreed price by a specified date. This approach: avoids the fragmented set of relationships which traditional construction requires customers to accept; provides a basis for effective communications with the customer; promises greater certainty in terms of quality, time and cost; and reduces the risks of customers being drawn into the conflicts, disputes and litigation which all too often characterise traditional construction.

Design build is particularly attractive to customers needing a straightforward building or simple infrastructure. Many of these customers want to describe their requirements to a construction company and leave them to produce the new facility at a sensible cost and time. The resulting relationship between customers and design build companies is commonly characterised as providing a single-point of responsibility.

In addition to providing a single-point of responsibility, the obvious strength of design build compared with traditional construction is it provides a straightforward way of ensuring practical knowledge of procurement, manufacturing, production and commissioning is taken into account in the brief, design and plan. This is sufficient advantage to ensure design build frequently delivers greater efficiency and more reliable quality in less time than traditional construction.

Design build is well established in civil engineering projects where the common professional background of customers' advisors, designers and contractors undoubtedly fosters this arrangement. Even when a traditional approach is used, there is a long history of civil engineering contractors submitting alternative designs as part of their bids. It is therefore a relatively small step for contractors to accept the larger responsibilities of design build.

The use of design build in building projects is well established in some countries and indeed has become the dominant approach. The most highly developed forms are described in Chapter 10. However, more basic approaches to design build demonstrate some of the construction management propositions and therefore form the subject of this chapter.

Construction Management Strategies: A Theory of Construction Management,
First Edition. Milan Radosavljevic and John Bennett.
© 2012 John Wiley & Sons, Ltd. Published 2012 by John Wiley & Sons, Ltd.

The effect of design build has been well researched in United Kingdom building projects. In this context diverse professional backgrounds create barriers to design build and highlight the changes required to move away from traditional construction. Significant changes occurred in the last decades of the twentieth century. Up to this time, the dominant role of architects allowed the Royal Institute of British Architects (RIBA) to retain rules which prevented architects from working in responsible positions in any form of contracting company. This rule was intended to emphasise the professional nature of architectural activities. It meant architects were unable to take the lead in design build companies unless they resigned from their professional body. Since many customers naturally sought advice from architects when they first considered the need for a new building, they were kept in ignorance of design build. The established relationships in traditional construction also inhibited general contractors from suggesting alternatives to architects' designs. Despite these barriers, design build has become significant in United Kingdom building projects.

6.2 Design build customers

The growth in the use of design build in the United Kingdom for building projects was led by developers and institutional investors who wanted relatively straightforward factories, warehouses and office buildings as commercial products. Provided they could find buyers or tenants, they did not have any further interest in their buildings. The demands of financial institutions were more important than the needs of occupiers and users. Cost certainty, speed and low exposure to disputes and litigation were sufficient to persuade these customers to use design build.

Seeing the success of developers and institutional investors using design build persuaded other customers to experiment with the approach. It was these pressures which forced RIBA to change its rules about architects working in contracting companies. It also led the Joint Contracts Tribunal (JCT), who publishes the most widely used contracts for building projects in the United Kingdom, to introduce a With Contractor's Design form in 1981. This was a hastily modified version of the traditional contract between customers and general contractors. This gave customers a widely accepted basis for contracts with design build companies. The initial form has been revised and is now published as the well-thought-out JCT Design and Build Contract. The approach has taken its place in the construction industry's normal ways of working alongside traditional general contracting (described in Chapter 5) and management approaches (described in Chapter 7).

The early use of design build was messy and involved various compromises before the best approaches emerged. The most straightforward approach is for the customer, with advice from architects or quantity surveyors, to appoint a design build company at the inception of a project as illustrated in Figure 6.1. This is usually achieved by negotiating with one or, at the most, three design build companies.

A widely-used approach is for the customer to employ design consultants to produce the brief and design. This is developed to a stage which enables the customer to understand their new building and be given realistic estimates of the cost and completion date. Then a general contractor is selected usually on

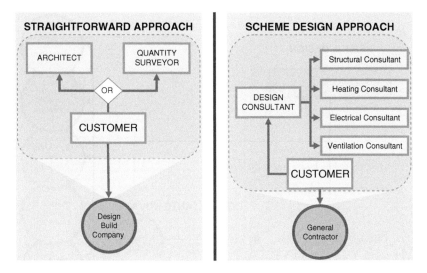

Figure 6.1 Common Approaches to Design Build.

the basis of competitive bids as illustrated in Figure 6.1. The selected general contractor is required to take over the employment of the customer's consultants. This is called consultant novation because the employment contracts of the consultants are novated to the general contractor. In this way parts of the general contractor's organization act as a design build company for the duration of the project. This temporary design build company takes responsibility for the design whatever stage it has reached when they are appointed. Then it provides the customer with a single-point of responsibility for the completion of the project. The consultants work with the design build company to complete the design details as illustrated in Figure 6.2. The outcomes tend to be dogged by muddled responsibilities and generally provide poor outcomes for everyone involved.

A more effective approach which enables customers to use consultants to produce a brief and undertake some design is a variation of design build which is often called develop and construct.

As shown in Figure 6.3 the customer employs consultants to help them decide exactly what kind of building they need. Then a design build company is employed on the basis of negotiation or competitive bids and takes direct responsibility for completing the brief and design and undertaking all subsequent stages of the project. The design build company does not take over the employment of the customer's consultants. They use either directly employed staff or their own choice of design consultants and subcontractors in undertaking the project in accordance with the information provided by the customer for an agreed cost by an agreed completion date. The earlier the design build contractor is employed in the production of the brief and design, the more likely it is the project will be successful. The customer often retains some of their own consultants to ensure the design build company meets their contractual commitments. Apart from this supervision, which is usually minimal, the design build company is able to use their knowledge of procurement, manufacturing,

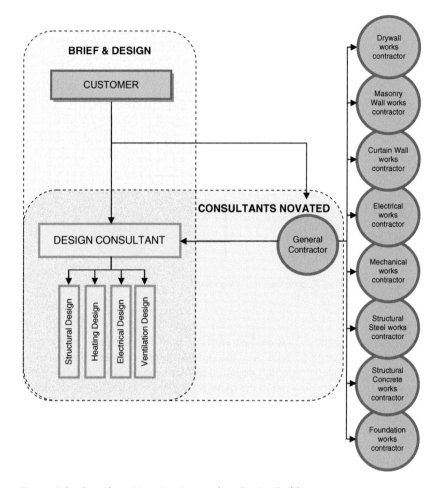

BRIEF & DESIGN

CUSTOMER

CONSULTANTS NOVATED

DESIGN CONSULTANT

General
Contractor

Structural Design

Heating Design

Electrical Design

Ventilation Design

Drywall
works
contractor

Masonry
Wall works
contractor

Curtain Wall
works
contractor

Electrical
works
contractor

Mechanical
works
contractor

Structural
Steel works
contractor

Structural
Concrete
works
contractor

Foundation
works
contractor

Figure 6.2   Consultant Novation Approach to Design Build.

production and commissioning to ensure the design and plan provide the basis for efficient work.

6.3 Design build companies

Pressure from customers wanting to use design build has resulted in some general contractors offering design build for building projects. Also a few design build companies have been established by architects. This is a recent phenomenon because of the changes in RIBA rules which meant architects wishing to be involved in senior roles in design build companies no longer have to resign from their professional institute. As a result some of the most highly regarded and effective design build companies are led by architects.

Whatever their origins, design build companies employ all the knowledge and skills required to undertake construction projects. They do this by some combination of directly employed people and subcontracting. Decisions about which knowledge and skills are provided directly by the company and which can

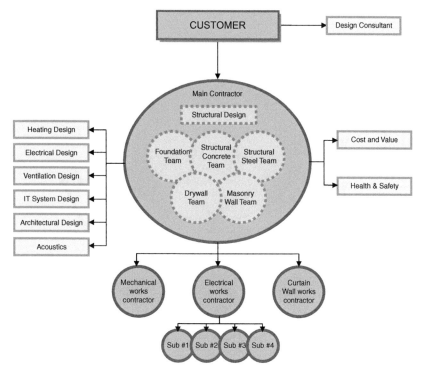

Figure 6.3 Develop and Construct Approach to Design Build.

be subcontracted are based on evaluating the risks, effectiveness, market conditions and many other factors. The resulting decisions affect the kind and quality of service design build companies are able to provide for customers and their performance and profitability as is described in the section 'Design build performance'.

In general design build companies specialise in related sets of construction technologies. The choice is dictated by local demand for particular types of facilities, and the experience and preferences of the company's directors. As the experience of the company and their subcontractors grows, construction teams become more experienced in ensuring designs and plans create a basis for subsequent stages to be undertaken efficiently. Construction teams become more knowledgeable and skilled in using the selected technologies and better at working together. All these benefits follow from the closely coordinated production of designs and plans by construction teams experienced in procurement, manufacturing, production and commissioning.

The project organization which puts design build into practice is influenced by the approach adopted by the customer and design build company. A typical pattern of boundary and internal relationships is shown in Figure 6.4 which illustrates a straightforward form of design build organization.

In the design build approach illustrated in Figure 6.4 it is relatively straightforward to internalise many of the boundary relationships which exist between contractually employed construction teams in developed traditional construction. Many design build contractors therefore develop internal specialist

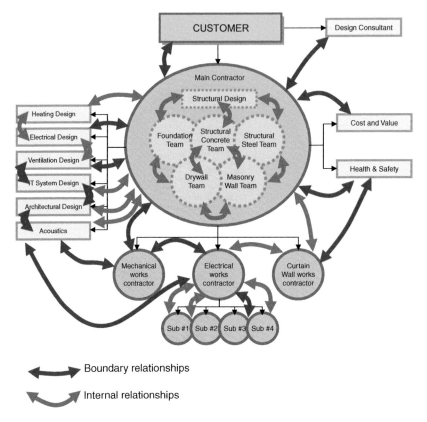

CUSTOMER

Design Consultant

Main Contractor

Structural Design

Heating Design

Electrical Design

Ventilation Design

IT System Design

Architectural Design

Acoustics

Foundation
Team

Structural
Concrete
Team

Structural
Steel Team

Drywall
Team

Masonry
Wall Team

Cost and Value

Health & Safety

Mechanical
works
contractor

Electrical
works
contractor

Curtain
Wall works
contractor

Sub #1 Sub #2 Sub #3 Sub #4

Boundary relationships

Internal relationships

Figure 6.4   Design Build Project Organisation.

teams that are brought into individual project organizations thus avoiding the potential costs and risks associated with contractual relationships. It is likely the internal teams know each other well because they have worked together on other projects. However, the downside of such an approach is it requires a relatively big organization, which in itself is likely to pose difficult organizational challenges, particularly in a rapidly changing and increasingly globalised market.

Many large general contractors seek to internalise relationships by resorting to mergers with or acquisitions of specialist contractors. While this eliminates the potential costs and risks associated with contractual relationships, it does not necessarily build effective internal relationships. Many mergers and acquisitions introduce insuperable cultural differences, which all too easily escalate into open confrontation between teams. As a result, what on the surface might look like a single organizational entity is often divided by cultural differences which give rise to boundary relationships inside the merged organizations every bit as damaging and inefficient as those which exist between contractually linked external organizations. Indeed the situation may be worse because the lack of contractual obligations may well mean teams faced with problems have no adequate legal instruments to resolve their difficulties. Problems escalate and projects are faced with highly detrimental situations. For these reasons, most successful design build contractors emerge as a result of organic growth. This may be supported by the occasional acquisition of comparatively small specialist contractors. When this is necessary, effective contractors invest in achieving a smooth blend of cultures so the new members of the organization quickly develop internal relationships with their new colleagues.

Another common modification of the straightforward design build approach is to have most of the major packages undertaken by experienced subcontractors. All too often this approach provides only the superficial appearance of a single-point of responsibility. Customers assume the design build contractor provides a clear single-point of responsibility for delivering the project but clashes between the contractor and powerful subcontractors can take the customer by surprise as the project is beset by costly disputes and substantial delays. These problems are best avoided by design build companies taking direct responsibility for all major packages and using sub-contractors only for relatively minor packages where any problems can be easily resolved. Then even if a dispute makes it necessary to replace a sub-contracting company, this can be achieved without major difficulties for the project or the customer.

6.4 Design build process

The most straightforward form of design build begins with the customer and a design build company discussing the new facility. The customer has more or less well developed ideas of their requirements and the design build company is competent to discuss various types of the required facility, examine the implications of different performance standards, explore alternative design ideas, as well as take account of the cost and time implications of the alter-natives. More than this the customer knows whatever is discussed can be delivered by the design build company. So if an agreement is reached, the customer knows what they will get, when it will be ready for them to occupy, and the cost. All this results from discussions with one design build company. In this way design build insulates the customer from the internal complexity of construction by providing a single point of responsibility for delivering the completed new facility.

Consultants' involvement

It is more usual for customers to employ consultants to help prepare a description of their requirements before they involve a design build company. This may be a simple statement of needs, a well developed brief, a concept design or a more fully developed design. These different bases involve designers in the project in different ways and involve the design build company at different stages in the overall project process.

Another important variable in the way design build projects are set up is the role of the customer's consultants after the design build company is given responsibility for constructing the required facility. The consultants may be retained by the customer to supervise the completion of the design and the subsequent actions by the design build company. Alternatively, the em-ployment of the consultants may be novated to the design build company.

As described earlier in this chapter, in the novated approach, the consultants become part of the design build company's project organization.

Selection processes

Another variable which affects design build projects is the way the customer decides to select a design build company. The most straightforward approach is where customers are able to select a design build company on the basis of previous experience of working with the company. Alternatively they may rely on recommendations from colleagues or friends or the design build company's reputation. These direct approaches allow the design build company to be involved in the project from the very first stage.

Many customers prefer to select a design build contractor on the basis of competition. This normally requires the customer to employ construction designers and some kind of managers to prepare the basis for competitive bids. In practice there are many different arrangements but they all seek to provide a description of the required facility, a basis for determining the cost and completion date. The more certain the customer is in describing their requirements, the more certain the design build company can be about the cost and time.

Design build organization

Once they have agreed a contract with the customer, on whatever basis, the design build company sets up: a design team to organize the completion of the brief and design; and a management team to produce a plan for procurement, manufacturing, production and commissioning, and actively manage the costs and time. Ideally the two teams work together, usually led by a project manager who ensures the teams cooperate in meeting the customer's requirements as efficiently and profitably as possible. Specialised knowledge and skills not available inside the design build company are provided by subcontractors employed as and when they are needed.

Communication with the customer

The project manager provides the main point of contact with the customer. Effective project managers give their customers every opportunity of being involved in decisions which affect the performance or standard of the new facility, or threaten to change the cost or completion date. They ensure they are readily available to discuss any concerns the customer may have at any stage of the project; and they deal with any of the concerns which prove to have substance immediately.

Brief, design and plan

The first actions undertaken by the design build company's internal team depend on the description of the customer's requirements available when they start their work. Their first aim is to ensure there is a complete brief which the customer understands and agrees. This may exist when they start their work but if it needs further work, the designers and managers will work with the customer to produce the brief. Then they concentrate on the design and plan. Design build enables this stage to take account of all the procurement, manufacturing, production and commissioning implications of design decisions more thoroughly than can be achieved by traditional construction. The aims at the design and plan stage are to produce a design which meets all the customer's requirements, which the customer accepts, and ensure the subsequent actions can be undertaken efficiently and profitably.

Procurement, manufacture, production and commissioning

Once an agreed design and plan exist, the design build company concentrates on carrying out the subsequent stages as efficiently and quickly as makes commercial sense. They use directly employed production teams where these exist but it is more common for most of the subsequent actions to be undertaken by subcontractors. Their work is coordinated by the design build company's project manager assisted by the design and management teams. Most forms of contract in common use clearly establish the responsibility of the design build company to satisfy the customer's requirements as defined in the contract or as subsequently amended by the customer.

Supervision by consultants

The customer may continue to employ consultant architects and quantity surveyors to ensure the design build company's proposals fully meet their requirements, check the quality, and agree the costs. However, some experienced customers provide minimal supervision of this kind and rely on the design build company to meet its contractual obligations.

6.5 Design build performance

There was a rapid growth in the use of design build for building projects in the United Kingdom during the 1980s and 1990s. This led research to measure the performance of the increasingly widely-used approach. Important results show design build has distinct advantages over traditional construction. It also shows performance is directly influenced by the approach adopted by design build companies.

The production and commissioning of new facilities when design build is used takes 12 per cent less time than similar projects using traditional construction. However, overall design build projects are completed some 30 per cent faster than similar projects using traditional construction. This illustrates design build's greater efficiency in decision making and communication at all stages of projects. This is particularly evident throughout the actions needed to produce and agree the brief, design and plan which are dominated by decision making and communication.

Design build projects are completed on or before the agreed completion date more reliably than traditional construction. This is achieved with greater

Process	% of projects completed on time or early
Design build company involved at the start of the project	78
Design build company involved after brief is prepared and scheme designs produced	70
Design build company involved after design is substantially complete	66
Traditional construction	56

Table 6.1 Relationship Between Project Approach and Time Performance

Source: Bennett et al. (1996)

certainty the earlier in the project process the design build company is involved. Table 6.1 shows this important effect.

Design build project costs are at least 13 per cent lower than similar projects using traditional construction. Also design build projects are completed at a cost which is reasonably close to the agreed budget more reliably than traditional construction. This is achieved with most certainty when the design build company is appointed on the basis of a substantially complete design which they take responsibility for completing. Table 6.2 shows this effect.

Design build projects have a mixed performance in terms of quality. Customers using design build tend to expect and generally get lower quality performance than those using traditional construction. However, customers' expectations about quality are most likely to be met when the design build company is involved from the start of the project. This approach outperforms traditional construction as Table 6.3 shows. This suggests quality is highest on projects where designers are employed on a consistent basis throughout the design processes. When responsibility for design is passed from consultants to a design build company, quality suffers. The worst situation is where the employment of the customer's design team is novated to the design build company. The customer's quality expectations were met on only 28 per cent of project which used novation.

Process	% of projects completed within 5% of the agreed budget
Design build company involved at the start of the project	76
Design build company involved after brief is prepared and scheme designs produced	69
Design build company involved after design is substantially complete	90
Traditional construction	68

Table 6.2 Relationship Between Project Approach and Cost Performance

Source: Bennett et al. (1996)

Process	% of projects meeting the customer's quality expectations
Design build company involved at the start of the project	68
Design build company involved after brief is prepared and scheme designs produced	48
Design build company involved after design is substantially complete	38
Traditional construction	60

Table 6.3 Relationship between Project Approach and Quality Performance
Source: Bennett et al. (1996)

A further significant result is the performance described in Table 6.3 is directly correlated to the level of quality expected by the customer. In other words, the higher the quality targets set by the customer, the greater the probability they will be met. This reflects research in other industries and it is significant that high targets for quality motivate construction teams to deliver high quality.

These results tell us that when a design build company's designers are responsible for the total design process, design build out performs traditional construction in terms of time, cost and quality. Indeed design build achieves a superior performance in terms of time and cost even when develop and construct, or novation is used. However, these two less straightforward approaches to design build are inferior to traditional construction in terms of quality. It suggests changing responsibility for design part way through the overall design process is inefficient in terms of quality.

Overall, these results show the greatest efficiency results from design build companies having total responsibility from the start of projects. This allows them to act on the basis of many of the construction management proposi- tions stated in Chapter 4. The benefits are reduced when design build companies are appointed later in projects and in terms of quality, perfor- mance is worse than traditional construction. As Figure 6.5 shows design build companies generally need to give more attention to quality. This should include setting specific and unambiguous quality requirements through a series of quality management workshops with customers. This provides a robust starting point for quality management and should lead to better results.

A perhaps predictable finding is traditional construction, where an architect takes the lead, was found to be more likely than design build to produce outstanding architecture. However, the same research shows architect-led traditional construction is just as likely to produce buildings widely regarded as ugly. In other words traditional construction was seen to deliver variable performance in terms of architectural standards. A further entirely surprising result is the physical appearance of facilities produced by design build were judged by a large and diverse sample of people to be on average superior to those produced by traditional construction. In other words, although archi- tect-led traditional construction provided most examples of outstanding architecture, design build was judged to deliver more consistent architecture

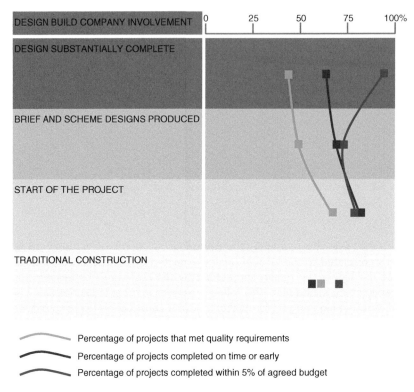

| DESIGN BUILD COMPANY INVOLVEMENT | 0 | 25 | 50 | 75 | 100% |

DESIGN SUBSTANTIALLY COMPLETE

BRIEF AND SCHEME DESIGNS PRODUCED

START OF THE PROJECT

TRADITIONAL CONSTRUCTION

⎯⎯⎯ Percentage of projects that met quality requirements

⎯⎯⎯ Percentage of projects completed on time or early

⎯⎯⎯ Percentage of projects completed within 5% of agreed budget

Figure 6.5   Design Build Performance.

than traditional construction. Any judgement about whether it is better to have all buildings providing consistently good architecture by using design build or a few outstanding buildings set in a generally more mediocre built environment by using traditional construction is an important matter for all local communities.

6.6 Design build efficiency

The previous section shows even the most basic forms of design build provide opportunities for construction firms to be more efficient than is possible with traditional construction. The benefits in greater efficiency emerge at all stages of projects but the basis for them is largely established as the design and plan are produced. The design build company's design and management teams can ensure designs and plans allow subsequent actions to be undertaken efficiently by organizing the actions of teams so they avoid interfering with each other. This means as far as possible providing clear and distinct work areas for construction teams which allow them to carry out their work in one continuous sequence. An essential part of this is checking that each construction team fully completes their work in the manner expected by subsequent construction teams. It also means production and commissioning teams in particular do not damage the completed work left by preceding teams.

The design build company's design and management teams should ensure the required actions are comfortably within the competence of readily available construction teams. This means producing predictable designs and plans and discussing them with the manufacturing, production and commissioning teams to ensure they are competent in the construction technologies being used.

The design build company's design and management teams can make sure the selected construction teams understand and accept the project objectives and they are employed on terms which motivate them to meet these objectives. This is most effectively achieved with directly employed teams and subcontractors who enjoy established relationships with the design build company. It also helps if the teams can be involved in the choice of projects undertaken and in selecting and developing the construction technologies the design build company specialises in using.

As projects get underway the design build company's design and management teams should routinely check that teams are working towards agreed objectives. Any failures to understand or work consistently towards achieving the agreed objectives need to be dealt with immediately. Ideally the teams directly involved will work out the best ways of re-aligning their work with the project's objectives. However, if problems persist, senior managers need to work with the teams to find robust answers.

These various initiatives all serve to reduce the need for construction teams to communicate with each other as they concentrate on ensuring their own work is completed efficiently. However, some communication between construction teams is inevitable and the design and management teams can ensure provisions are in place to ensure this essential communication is accurate. This means project organizations having effective management structures and communication systems and where necessary training the construction teams in their use. It also means checking that communications are effective. As with problems arising from a lack of focus on objectives, any failures of communication need to be dealt with immediately by the teams directly involved and senior managers should become involved only if the teams fail to find robust answers.

6.7 Construction management propositions

The actions described in the previous section are made possible by the advantages which design build provides compared with the constraints of traditional construction. These advantages are encapsulated in the following construction management propositions.

To enable construction to be undertaken efficiently within a design build approach, construction management aims to:

- Improve the quality of relationships between the construction teams involved in a construction project.
- Reduce the performance variability of the construction teams involved in a construction project.
- Select construction teams competent in the technologies required by the project in which they are involved.

- Ensure construction teams accept the agreed objectives of the project in which they are involved.
- Ensure construction teams are motivated to achieve the agreed objectives of the project in which they are involved.
- Foster accurate communication between the construction teams involved in a project organization.

6.8 The theory of construction management

This chapter provides a test of the theory of construction management. Design build creates opportunities for some of the construction management propositions to be acted on and enable construction projects to avoid some of the inherent weaknesses of traditional construction. Research shows design build projects achieve greater efficiency than traditional construction. This result is reinforced by the more detailed results which show the greater the opportunity for design build companies to move away from the methods of traditional construction, the greater the efficiency achieved. It can therefore be claimed the theory of construction management passes its first test; and the construction management propositions in the section 'Construction Management Propositions' are supported. This is reflected in the example project which follows. The example provides a further illustration of the use of the Inherent Difficulty Indicators.

Example – Building Project using Design Build

The example relates to new hotel project similar to that used as the basis of the example in Chapter 5 but in this case using a design build approach. The design build contractor is selected after an architectural consultant employed by the customer has produced a brief and concept design. There is some duplication of effort as designers employed by the design build contractor have to reconcile the customer's concept design with the design build contractors preferred design details. The benefits of having a design which fits the design build contractor's preferred ways of working provides minor reductions in the effort required to prepare the detail design and production information, plan the project and select the construction teams compared with the traditional construction approach. More substantial benefits are provided as the design build contractor's detailed planning, control, contract administration and payment systems are well understood by the companies providing production and commissioning teams. Similarly the use of well-developed design details provides substantial benefits in some production actions, particularly those concerned with the internal fitting out and decoration of the hotel. The overall effect is the design build approach is likely to be completed faster and cost less than the traditional construction approach. In the example project the customer's architectural consultant supervises the design build contractor's work to ensure it meets the customer's needs.

		Number of Teams	Number of team-days
Brief		1	80
Design	Concept design	1	100
Procurement	Select design build company	2	100
Design	Scheme design	3	150
	Detail design	2	60
	Technical design	6	150
	Production information	2	100
Plan		2	150
Procurement	Select companies	3	200
	Design build contractor plan and control	3	250
	Contracts and payment systems	1	50
Manufacturing		5	150
Production	Substructure	4	100
	Structure	7	200
	External envelope	6	250
	Service cores	6	150
	Risers and main plant	5	120
	Entrance and vertical circulation	12	450
	Internal divisions	3	120
Decoration		10	600
Fittings		6	400
Landscaping and external services		6	250
Designer's supervision		2	180
Commissioning		7	140
TOTALS		105	4,500

According to equation. 4.11 there are R = 5,460 possible relationships between teams working on this project. The detailed project plan, which can be seen on the website linked to this book, www.wiley.com/go/constructionmanagementstrategies, shows 3,687 of the possible relationships do not occur. This gives a total of R-k = 1773 relationships which do in fact occur.

The largest number of teams which could be involved in the same time interval is 29 in the intervals 22 and 32 so there could be 406 different relationships within each of the two time intervals. However, since not all teams work at the same time it is imperative to determine the number of teams which work simultaneously for each time interval. In this case the plan shows that during the 32nd time interval the largest number of teams work simultaneously:

- 4 out of 6 external envelope teams;
- 4 out of 6 service core teams;
- 5 out of 5 risers and main plant teams;
- 6 out of 12 entrance and vertical circulation teams;
- 1 out of 1 contracts and payment systems teams; and
- out of 3 design build contractor's plan and control teams.

This gives a total of 22 teams with 231 relationships.

Every team on a project has established relationships with at least one but often several other teams giving a total of 1,128 established relationships. This gives an Established Relationships Indicator: $E_R = 1128/1773 = 0.64$ (equation 4:13). This is a reasonably good value which is then further confirmed by equation 4.14 which generates time dependent indicators of established relationships ranging from 0.38 to 1.00.

The relationship fluctuation indicator F_E is calculated using equation 4.15 and gives a value of 0.75. This reflects a strong influence of established relationships throughout the project because both E_R and F_E are comparatively high (see the example in Chapter 5).

However, to really establish the benefits of established relationships we have to look at how long the teams involved are working together on this project and how much time they have spent working together on past projects. Then equation 4:16 gives a Relationship Quality Indicator for each established relationship which ranges from 0.35 to 0.98 and the total Relationship Quality Indicator (equation 4:17) is: $Q_R = 834/1773 = 0.47$. Although there are a number of high values for individual Relationship Quality Indicators, the total Relationship Quality Indicator is only moderately good because the 1,128 established relationships out of possible 1,773 relationships are of moderate quality.

According to equation 4.18 we can calculate the relationships configuration indicator C_R:

$$C_R = \frac{1}{3}\left(\left(1 - \frac{34}{209}\right) + \left(1 - \frac{58.53}{1980}\right) + \left(1 - \frac{22}{105}\right)\right) = 0.87$$

The indicator shows the project organization is not overly complex. There are 34 time intervals but there could be a maximum of 209 time intervals. The maximum number of teams working at any given time is 22 in the 32nd time interval. The intervals are relatively short in comparison to total project duration, which is beneficial in order to prevent escalation of problems when teams interact for prolonged periods.

The past performance of the 105 teams over their immediately previous 10 projects provides individual team Performance Variability Indicators which range from 0.30 to 1.0. These give a total Performance Variability Indicator (equation 4.20): $R_p = 0.78$.

This is of course the average performance variability showing the teams managed to complete their work within the agreed time on 78% of their recent projects (note we have assumed c_j to be 1 for all teams although for real life projects these values should be determined separately).

The External Interference Indicator based on historical records for the particular region for the same type of building project is 0.45.

Overall, we have the following inherent difficulty indicators which are compared to the indicators for the similar project using traditional construction:

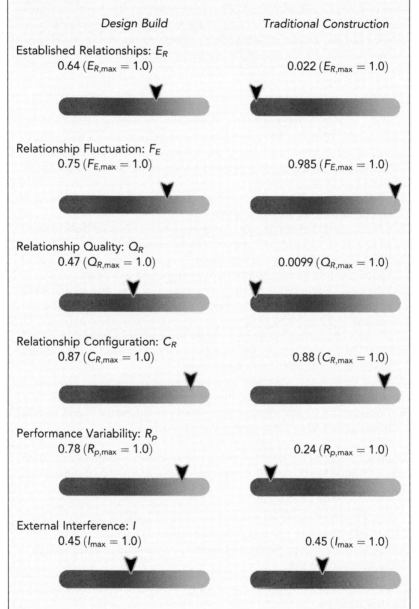

Design Build	Traditional Construction

Established Relationships: E_R
0.64 ($E_{R,max} = 1.0$)   0.022 ($E_{R,max} = 1.0$)

Relationship Fluctuation: F_E
0.75 ($F_{E,max} = 1.0$)   0.985 ($F_{E,max} = 1.0$)

Relationship Quality: Q_R
0.47 ($Q_{R,max} = 1.0$)   0.0099 ($Q_{R,max} = 1.0$)

Relationship Configuration: C_R
0.87 ($C_{R,max} = 1.0$)   0.88 ($C_{R,max} = 1.0$)

Performance Variability: R_p
0.78 ($R_{p,max} = 1.0$)   0.24 ($R_{p,max} = 1.0$)

External Interference: I
0.45 ($I_{max} = 1.0$)   0.45 ($I_{max} = 1.0$)

The design build project's indicators are reasonably high for Established Relationships, Relationship Quality, Performance Variability and External Interference. These indicators suggest the project has a manageable level of inherent difficulty. This is reinforced by the reasonably high values of the indicators for Relationship Fluctuation and Configuration which show the reasonably good quality of relationships in an appropriately optimised configuration tend to benefit most of the project. There is therefore a reasonable probability the project will be completed much as planned.

This reflects the teams' reasonably good performance resulting from a relatively large number of established relationships which benefit a significant proportion of the interactions. There is a likelihood of external interference but the project organization's decision to use design build gives them a realistic chance of completing the project as planned.

Exercise

Consider the 10-storey 300-room 5-star hotel project using a design build approach which forms the basis of the Example given in this chapter.

Consider this and provide a detailed answer to the following questions:

1. Who should be involved in producing the brief ? Explain the level of design detail they should produce.
2. What is the role of design consultants and what options are there in terms of how they will work with a design build company?
3. What is the role of the design build company? Describe the use of subcontractors as opposed to directly employed workforce.
4. Why could competitive bidding as a means of selecting design build companies lead to problems during the detailed design and subsequent stages?
5. What should be the first aim in using the design build approach and why?
6. How would you assure the quality of the delivered facility meets customer expectations and requirements?
7. What steps would you take to ensure construction teams involved in design build projects are capable of meeting project objectives?
8. Consider the traditional approach described in Chapter 5 and identify how it differs from the design build approach.

Consider the following situation which may arise during construction. Think why a situation like this can be detrimental and what could be done to avoid such problems in a project.

A large subcontractor responsible for a major package on the critical path disagrees with the design build contractor's payment regime and resorts to the drastic measure of sending only a handful of workers to the construction site.

Further Reading

The following publications are the source of ideas used in this chapter and provide further information for readers.

Bennett, J., Pothecary, E. and Robinson, G. (1996) *Designing and Building a World-class Industry*. Centre for Strategic Studies in Construction, Reading. This report describes the results of extensive research into the performance of design build. It also provides recommendations for practitioners.

Pain, J. and Bennett, J. (1988) JCT with Contractor's Design form of contract: a study in use. *Construction Management and Economics*, 6 (4), 307–37. This paper describes research into the use of the JCT with Contractor's Design form of contract.

Thomas, A. (2006) *Design-Build (Architecture in Practice)*. Wiley-Academy. The book provides a straightforward and easily followed review of various aspects of design build, from procurement to completion. It is illustrated by descriptions of recent design build projects in the United Kingdom.

Chapter Seven
Management Approaches

7.1 Introduction

Throughout the 1970s customers responsible for major commercial developments became increasingly frustrated by the slow, unreliable performance delivered by traditional construction. One response was to bring professional management teams into construction projects. This occurred first in the United States in the 1970s and 1980s where construction management rapidly established itself as an effective approach for large, complex building projects. A combination of highly commercial customers and experienced consultants and contractors, particularly in such major cities as New York and Chicago, led to an approach which is generally called construction management.

Various ways of involving management were tried but they all included a management team working in cooperation with a design team under the general direction of the customer. This triumvirate produced the brief, design and plan and procured specialist construction companies to undertake the manufacturing, production and commissioning as shown in Figure 7.1.

The approach was successful in producing outstanding buildings significantly faster and more reliably than traditional construction. It suited developers to begin earning a return on their massive investments at the earliest possible date. This was more important in securing the financial success of development projects than lower construction costs.

Construction management companies developed rapidly as major commercial customers employed them on their new developments. The success of construction management was highly publicised in the construction industry's media. Researchers from many countries visited the United States to study the new approach to building projects. Their reports, which recommend the use of construction management, attracted the attention of customers. This was particularly the case in London where in 1986 the Thatcher government deregulated the financial markets in the so called 'Big Bang.' This led to a massive demand for new high quality offices as London emerged as the world's most important financial centre.

The opportunities in London attracted the attention of the leading United States construction management companies. Several of them set up new business expressly to bring their new approach to the United Kingdom. They quickly discovered United Kingdom architects worked in a very different way from that which they were used to. The established United States approach to construction management did not work with United Kingdom architects and a

Construction Management Strategies: A Theory of Construction Management,
First Edition. Milan Radosavljevic and John Bennett.
© 2012 John Wiley & Sons, Ltd. Published 2012 by John Wiley & Sons, Ltd.

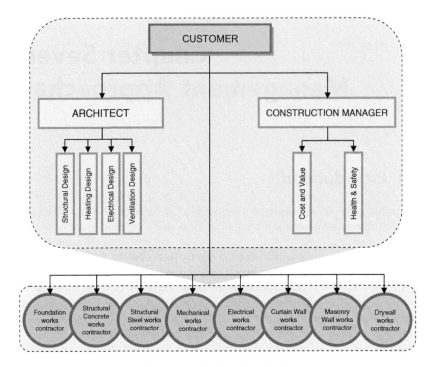

Figure 7.1 Construction Management Project Organisation.

number of projects ran into difficulties. These had a number of causes, including designs which challenged the capabilities of manufacturing and production companies and United States management methods which lacked the flexibility to deal with design changes, delays and claims for extra costs. Most of the American construction management companies closed their United Kingdom businesses. However, a few recognised the strengths of United Kingdom architecture and worked with leading customers to develop what some described as a mid-Atlantic approach. Also a few leading United Kingdom contractors recognised the strengths of the construction management approach and worked with leading customers to establish effective ways of bringing management into major building projects. Pre-eminent among these was Bovis who had used a management approach to contracting for several decades and recognised their time had arrived.

These diverse pressures and opportunities all came together in the mid 1980s at Broadgate in the City of London. This major development, comprising 14 outstandingly successful building projects and the creation of two major public open spaces, used construction management. The Broadgate approach to construction management was developed by the customer, Rosehaugh Stanhope, the United Kingdom management contractor, Bovis, the United States construction management company, Schal, the United Kingdom designers, Arup Associates, and the United States architects, Skidmore, Owings and Merrill. It brought together a team of outstanding people from both sides of the Atlantic and provided an outstanding demonstration of the level of efficiency construction management can deliver. The following description of construction management is largely based on the Broadgate approach because it exhibits most of the key features of best practice.

7.2 Customers

Construction management requires an experienced customer. It suits the production of large, city centre buildings particularly where the customer wants high-quality architecture. It has also been used successfully for major out-of-town shopping centres, for example, the massive Bluewater situated just off the M25 motorway which circles London. In other words, construction management suits major commercial developers, particularly those able to be involved in detail in their construction projects.

Construction management requires the customer to provide leadership in establishing the overall objectives for projects, selecting companies to provide the design and management teams, and arranging the finance. The most effective also establish clear guidelines for the way everyone involved will work together. They determine the size, shape and configuration of the buildings they want and go further by specifying the technologies designers will use in their designs. They encourage the use of large, prefabricated components wherever they serve to enable fast production. They establish the nature and methods to be used in planning and controlling quality, time and cost. They expect everyone involved to be involved in setting the targets which guide their work and to meet those targets exactly. They set out in great detail exactly how production on site will be organized and place direct responsibility on all the companies they employ to put their approach into practice fully and efficiently. They define the training to be given to all production and commissioning teams when they first arrive on site. They specify exactly how and when the building will be completed and handed over to them.

Having provided clear leadership at the start of their project, the customer is actively involved in all the subsequent stages. Their involvement is established in the contracts between the customer and design team, management team and works contractors who undertake the manufacturing, production and commissioning. The customer is represented at project meetings where: they ensure everyone is focussed on the project objectives; problems are solved fully and quickly; and good ideas are discussed and either used to benefit the current project or recorded for future use. Their final task is to accept the building on completion and assume responsibility for it. That is their final task unless there are defects in the finished building, in which case the customer ensures those contractually responsible remedy the problems.

7.3 Designers

The design team in building projects is led by an experienced architect who works with the customer and construction manager in producing the brief. The lead architect is primarily responsible for determining the look and feel of the building. In the United States approach this means leaving much of the detail design to specialist manufacturing and production companies. In the United Kingdom approach the architect remains involved and responsible for the detail design as illustrated in Figure 7.2. The result is the best United Kingdom buildings have a real depth of quality which qualifies them as great architecture.

Much of the detail design is undertaken by engineers. Some are consultants responsible for the overall design of foundations and structure. Other

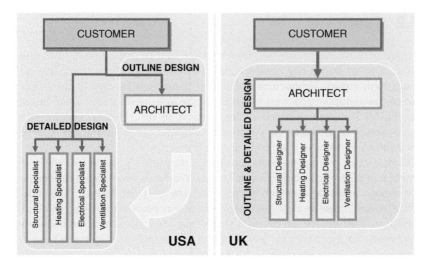

Figure 7.2   United States and United Kingdom Approaches to Design in Construction Management.

engineering consultants are concerned with safety, environmental impact, internal and external services, acoustic properties and all the other technical aspects of buildings. Many engineers involved in the design of buildings are employed by companies responsible for manufacturing major components. These are likely to provide the structural frame, external cladding, floors, internal partitions, major elements of internal services and even complete internal spaces such as toilets, bathrooms, plant rooms and other heavily serviced spaces.

The architect leading the design team is responsible for coordinating the work of all the various designers to ensure the design is complete and fully meets the customer's requirements. They ensure design information required by statutory public authorities is ready and in the correct form when it is needed. They work with the customer and construction manager to ensure all necessary approvals are obtained. They also ensure the construction teams involved in the project are not delayed because design information is late or includes errors.

These responsibilities require the lead designer to: establish clear stages in the design process; organize design meetings which bring together all the people with an interest in the current design stage; provide the meetings with everything they need to make properly considered decisions; ensure the required design information is complete at the end of each design stage; check manufacturing and production teams understand the design information they are required to work with; and ensure design information is not changed once the design stage which produced it has ended.

The design team cooperate with the customer and construction manager in establishing quality control systems to ensure components are manufactured, and production on site is carried out in accordance with the design. The quality control systems specify the tests and inspections which must be carried out and define the records which must be kept of the results. The systems particularly ensure commissioning is undertaken thoroughly so the finished building functions fully in accordance with the design.

7.4 Construction managers

The construction management team is led by an experienced construction manager who works with the customer and lead designer in producing the brief. The construction management team ensures the building described in the brief can be constructed by the agreed date within the budget. As part of this initial judgement, they ensure the site is thoroughly evaluated to identify constraints and opportunities which may affect the brief and any subsequent stages of the project.

The construction management team works with the customer and designers to ensure designs can be manufactured, produced and commissioned efficiently at times and costs which meet the customer's requirements. They try to reduce the number of separate works contractors needed by selecting technologies which provide major elements of the finished building and looking for opportunities to use large prefabricated elements which simplify and reduce production on site.

The construction management team takes the lead in designing the project organization in a manner which establishes clear roles and responsibilities. As part of this they plan the time and cost available to each construction team at each stage of the project. This guides the procurement of specialist companies able to supply competent individuals and teams to undertake each role. Procurement is primarily the responsibility of the construction management team but it requires the close cooperation of the customer and lead designer to ensure the selected companies fully understand and accept their responsibilities and how these relate to the rest of the project organization.

The construction management team ensures the contracts which the customer enters into with specialist construction companies are properly formed. This is a large task since a major building project may well involve the customer in a hundred or more individual contracts. The customer will almost certainly have expert assistance from lawyers but the construction management team needs to ensure the contracts are consistent with each other and in total cover all the manufacturing, production and commissioning activities.

The construction management team establish the systems needed to ensure the required building is complete by the agreed date within the budget. Effective systems begin by involving designers, managers and works contractors in establishing the detailed cost and time targets which apply to their work. This needs to be genuine involvement so everyone constantly strives to meet their targets. As production work progresses, the construction management team ensure works contractors stick to the overall project plan. They set up control systems to monitor time and cost and identify any potential problems at the earliest possible stage. This involves regular progress meetings which must be attended by a responsible representative of every company involved in the current stage of the project. The meetings ensure problems are dealt with immediately by those directly involved. When a progress meeting fails to agree the solution to a problem, the construction management team convenes a special meeting. These special meetings take priority over all other work and continue, if necessary over several days, until a solution is agreed. Problems are not allowed to fester and grow into crises which threaten the efficient construction of the building.

An important role of progress meetings is to identify good ideas. These may result from designers tackling an unusual situation, managers finding a better way of working or a works contractor finding a way of reducing the time or cost of their work or seeing a way of improving quality. Experienced construction managers know good ideas may disrupt progress on the current project. They therefore ensure ideas are evaluated and a mature decision made about using it on the current project or recording it so it is considered for future projects.

The construction management team ensures formal contractual documents dealing with changes and payments are dealt with properly in ways which avoid delays or disputes. They ensure notices and certificates required to comply with regulations and legislation are issued and received. They ensure the building users' manual is prepared as each stage of the project is complete so when it is handed over to the customer it provides a clear and complete guide to the efficient and effective use of the new facility.

7.5 Works contractors

Manufacturing, production and commissioning are carried out by works contractors. The term works contractors, rather than subcontractors, is used because in the construction management approach the companies have direct contracts with the customer.

Some works contractors form part of the design team by providing specialist technical knowledge and ensuring their elements of the building are capable of being manufactured and produced efficiently. This normally includes producing shop drawings used in manufacturing and may include production drawings for specialised work. Throughout their involvement in design, works contractors ensure the cost and time targets agreed for their work are capable of being achieved.

Works contractors agree with the construction management team how and when their manufacturing, production and commissioning actions will be carried out. In particular they agree how their work on site will be organized and coordinated with the work of other construction teams. These decisions are recorded in a detailed plan for each works contractor's actions on site.

Works contractors manufacture components, obtain materials and ensure they are all delivered to site in accordance with the plan. They ensure their site based construction teams attend induction courses run by the construction management team so they know how the site is organized before they begin work. The site-based construction teams take control of their work areas, ensure everything they need is in place, contribute to progress meetings, solve any problems and ensure their work is fully complete in accordance with the plan. They carry out their work safely in accordance with all relevant regulations and legislation and leave their work area clean and tidy.

7.6 Construction management process

The construction management process is driven by the brief, design and plan. Where these are prepared thoroughly, the customer, design team, management team and the construction teams provided by the works contractors have every chance of undertaking their work efficiently and profitably. Achieving this requires the management systems to be used consistently and effectively throughout the project.

Brief

The key stages in construction management projects begin with the preparation and development of the brief. The customer needs to analyse their reasons for wanting a new building. They need to be clear what the building will provide in terms of the size and standards of accommodation and the services and equipment it will contain. They need to consider all its impacts on the local environment. They need to know where it will be constructed, and when it needs to be available. They need to undertake careful studies to ensure there is no better alternative, and be certain the necessary finance is available.

Customers may need to employ various kinds of expert advice before deciding to go ahead with the project. Once that decision is made, their first actions should be to select and employ a lead designer and a lead construction manager. They will join the customer in reviewing the brief, whatever stage it has reached, and completing it. It is likely the designer and construction manager will be able to suggest ideas not previously considered likely to give the customer a better building, quicker and at lower costs. These should be discussed and agreed decisions incorporated in the brief.

Design, plan and procurement

The next stage is intensively creative. The customer, designer and construction manager consider alternative design concepts and construction technologies. They may well draw on other opinions, ideas and expertise from within the customer's organization, the design company, the construction management company or possibly other organizations. These may include research institutes, government agencies, manufacturing companies and artists. The aim is to agree a scheme design and project plan which best meet the customer's requirements.

Once the scheme design and project plan are finalised and all the formal approvals are obtained, the project enters a hectic stage. The works contractors are selected, some on the basis of their reputations and others by competitive bids. Those responsible for foundations and basements, major structural elements, and elements which involve long manufacturing processes are selected first. Then all the others are selected in time to give them every opportunity of meeting their contractual obligations.

The first task for most works contractors is to produce the detail design for their work in cooperation with the design and construction management teams. In parallel with this the construction management team help works contractors plan their work in complete units which fit predetermined work areas on site. This begins with the overall plan which describes the broad sequence of work and the production methods to be used. Then manufacturing is planned in detail including drawings, manufacturing, quality control, storage and delivery to site. Finally the specific production actions in each works area are planned in terms of men, machines, materials and components including access, preparation and cleaning up once work in the area is complete. The detailed design and plan lead into the production of shop drawings for manufacturing, and production drawings for site activities.

Manufacturing, production and commissioning

The design and construction management teams authorise manufacturing work in time for the resulting components to be delivered to site in accordance with the project plan. The construction management team organize everything needed for production on site to begin. They provide fully-serviced office space for every company required to work on site. They arrange for a canteen, toilet and washroom facilities. The aim is that workers do not need to leave the site during the working day and they have no excuse for eating or urinating in their work places. The construction management team organizes the equipment and access needed for workers, materials and components to be moved to the correct work areas easily and quickly in accordance with the project plan. They plan how work areas and completed work will be protected from bad weather. They plan the site communications systems to allow rapid intercommunication between works contractors' site managers, the construction management team, design team and customer. They plan the collection and disposal of rubbish to ensure the site is kept tidy. They plan the provision of power, water and drainage to work areas. They plan rigorous grid lines and datums to ensure the building is produced accurately.

Production is undertaken in distinct work areas defined by the construction management team and clearly identified by physical barriers and markers. It is normal for structural work to be organized in work areas each containing one week's work for one works contractor. During later stages when production is more detailed and inter-related, each work area contains three days work. The works areas are designed so one works contractor can be given exclusive control for the one week or three days needed to complete their work.

Once there is a firm agreement about what they are to produce, the first works contractors set about producing the substructure. Other works contractors are employed in time for them to meet the demands of the project plan. Production of the building is undertaken in distinct stages which typically include: substructure; structure; external cladding; service cores, risers and main plant; entrance and vertical circulation spaces including finishes; internal divisions and finishes to horizontal spaces; and external works.

Works contractors appoint fully competent construction teams for each work area. The construction teams are responsible for controlling the activities in their work area. This means putting the detailed plan into effect so their work is fully complete in the agreed time. The construction team checks their work area is suitable for their work to go ahead. They ensure the boundary is clearly marked and access is permitted only to the construction team and members of the construction management team for setting out, inspection or reasons of safety. They check that all the plant, equipment, material and components are available or are due to be delivered in accordance with the detailed plan. They leave their work area clean with all the required work complete and tested by the end of the week or three days.

Typically construction teams are briefed about their day's work at the start of the day by the team leader. Any problems are solved by the construction team as they arise. All the construction team leaders working on site meet with members of the construction management team every day, usually in the early afternoon, to review progress and deal with any problems the teams have been unable to solve.

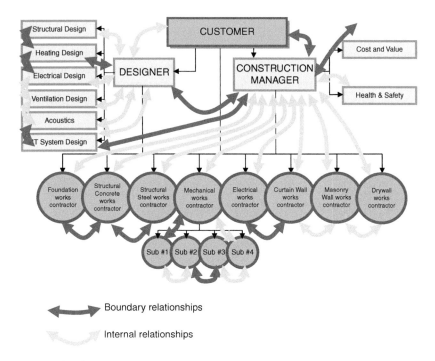

Figure 7.3 Construction Management Project Organisation.

Production and subsequently commissioning are driven forward by weekly progress meetings of senior managers of all the works contractors currently involved in the project, the customer, construction manager and designer. Each company reports they will meet all key dates or there is a problem preventing this. The meeting concentrates on solving any of these difficult problems. The success of construction management depends to a large extent on the weekly progress meetings being absolutely open and honest. In well-run projects, it is totally unacceptable to hide problems. In return individuals are never blamed for problems, everyone concentrates on finding solutions.

Hand-over

When production and commissioning are complete the building is handed over to the customer. They are provided with clear and comprehensive information on running and maintaining the building. Any defects are reported by the customer to the design and construction management teams who arrange for the responsible works contractors to remedy the problem.

The project organization which puts the construction management approach into practice is illustrated in Figure 7.3.

7.7 Construction management propositions

The actions described in the previous section aimed at improving the efficiency of construction management projects represent a highly-developed approach.

It is difficult to achieve and is largely dependent on customers with major construction programmes so manufacturing, production and commissioning teams become very competent at using the systems set up by the design and construction management teams. More generally the construction management approach creates a clear framework of systems and organizational arrangements which are encapsulated in the following propositions provided by the theory of construction management.

To enable construction to be undertaken efficiently, construction management aims to:

- Reduce the number of individual construction teams involved in a construction project;
- Improve the quality of relationships between the construction teams involved in a construction project;
- Reduce the performance variability of the construction teams involved in a construction project;
- Reduce the external interference experienced by the construction teams involved in a construction project;
- Select construction teams competent in the technologies required by the project in which they are involved;
- Ensure construction teams accept the agreed objectives of the project in which they are involved;
- Ensure construction teams are motivated to achieve the agreed objectives of the project in which they are involved;
- Foster accurate communication between the construction teams involved in a project organization; and
- Ensure construction teams regard the transactions which brought them into a project organization as advantageous to themselves.

7.8 Construction management performance

A distinctive characteristic of the construction management approach is all the key teams have direct contractual relationships with the customer rather than each other. This puts real pressure on construction management teams to establish close working relationships with design teams and works contractors if projects are to be successful. In considering construction management performance it is important to remember that serious problems during a project can trigger a breakdown of these relationships.

For example, late design changes requiring significant rework can lead to works contractors attempting to use their contractual relationships to deal directly with the customer rather than discuss their claims for extra time and money with the construction management team. In extreme cases this can lead to a complete communication breakdown in the project organization. The need for well-established relationships between construction management teams and works contractors over several successful previous projects is of paramount importance in avoiding these problems.

Creating an over-complex project organization to address the need for control is another potential pitfall of the management approach. Particularly on large-scale projects, a construction management company may wish to establish a very tight control of every aspect of works contractors' operation which results in a complex organization which includes several levels of management. Although originally designed to improve the control of a project, such an approach may obstruct progress by generating a large number of meetings many of which are needed simply to coordinate the various levels of management. This management complexity faced with problems, such as those caused by the late design changes described above, can seriously reduce the efficiency of everyone involved. Best practice relies much more on direct relationships between the construction teams which need to interact. This is commonly achieved by providing a joint coordination office for all works contractors on site to enable continuous interaction which simplifies communication and reduces the need for an excessive number of meetings.

Other problems may be caused if, due to an economic downturn or financial restrictions of some kind, competitive bidding is used to select works contractors. Where the lowest price bid in competition is allowed to take precedence over the quality of established relationships, there is an inevitable risk of bringing together a diverse selection of companies with considerably variation in their accepted standards of work. This can create difficulties in meeting project objectives and result in very slow and inefficient production. In such a situation, progress meetings are forced to focus primarily on rectifying misunderstandings about the required standards rather than solving work-related problems. Meetings all too easily produce crises which threaten the efficient construction of the building and require drastic measures which may be far more time consuming and expensive than the special meetings which are normally used to rectify serious problems on construction management projects. This all serves to emphasise that the construction management approach is heavily dependent on the ability of construction teams to work together in a coherent and well-established manner.

When it benefits from well-established relationships, construction management provides opportunities for construction firms to be more efficient than is possible with traditional construction. The benefits in greater efficiency emerge from the very start as customers set challenging objectives for their projects. They reinforce the good start by remaining closely involved in all key decisions, specifying the technologies to be used and ensuring they employ experienced companies to form their project organizations.

Customers work with their design and construction management teams to ensure designs and plans allow subsequent actions to be undertaken by the minimum number of construction teams consistent with efficiency. They combine or eliminate individual technologies by simplifying design details or prefabricating complex components.

The design and construction management teams involve works contractors in the design and planning of their work to ensure they are able to provide construction teams competent to undertake the production and commissioning actions. The construction management team provides induction courses for construction teams undertaking production and commissioning activities to ensure they understand how the site is organized. This all helps ensure construction teams understand and accept the project objectives. To ensure this is not undermined, the construction management team check construction teams are employed on terms which motivate them to meet these objectives.

The construction management team puts considerable effort into ensuring the actions of teams undertaking production and commissioning do not interfere with each other. They do this by defining work areas and ensuring teams can work independently and carry out their work in one continuous sequence. The construction management team insists on planning the delivery to site of materials and components in a manner which ensures work is not interrupted. They also ensure each team undertaking production and commissioning fully completes their work and leaves the work place in the manner expected by subsequent teams. They also arrange for completed work to be protected so teams do not have to return to the site to repair damage caused by subsequent activities.

Customers work with their design and construction management teams in establishing systems to check that all construction teams are working towards agreed objectives. Any failures to understand or work consistently towards achieving them are dealt with immediately by the teams directly involved. Where the problem persists, the customer, designer and construction manager involve senior people to help the teams find robust answers. This is all part of the relentless drive to ensure construction teams can work efficiently at meeting the project objectives.

Customers try to build long-term relationships with design and construction management companies and involve them in discussing future projects. Similarly they try to involve key works contractors in discussing and developing the construction technologies to be used on their projects. This helps ensure projects provide few surprises for those involved.

These various initiatives all serve to reduce the need for construction teams to communicate with each other as they concentrate on ensuring their own work within individual projects is completed efficiently. However, some communication between construction teams is inevitable and the design and management teams act to ensure this essential communication is accurate. This includes establishing communication systems and training everyone involved in their use. It means checking communications are effective. This is all embodied in a project culture which encourages any failures of communication to be dealt with immediately by the teams directly involved; senior managers are brought in only if the teams fail to find robust answers.

Given the clear opportunities to achieve high levels of efficiency which the construction management approach provides, it is curious there is little research into the performance of projects using the construction management approach. In part this is because the most publicised examples are projects for developers who wanted large, complex buildings constructed as quickly as possible. They also wanted certainty in respect of quality and cost but their primary aim was fast construction. There are few examples of comparable projects using traditional construction which makes it difficult to make rigorous comparisons of performance.

There is little doubt that construction management is considerably faster than traditional construction. Major developers in the United Kingdom justified their use of construction management in the 1980s by claiming their buildings were completed 6 to 12 months earlier than the two or more years it would have taken if they had used traditional construction. They clearly believed this because they began using construction management for their projects which involved investing hundreds of millions of dollars. Some data from this time is available from a careful comparison of performance achieved on large building projects in the United Kingdom, where traditional construction was the norm, and the United States which was using construction management provides some support for the developers' beliefs.

Different regulatory systems in the two countries made it impossible to make any direct comparison of the time between the customer agreeing a brief and the start of production on site. This was much quicker in the United States largely because delays in obtaining planning permission in the United Kingdom could amount to many years. The relative performance of production on site could be compared directly and showed that United States construction management enabled production to be about 40 per cent faster than United Kingdom traditional construction. The same studies showed building project costs were similar in the two countries.

7.9 Other management approaches

Best practice in the construction management approach, as described in this chapter, places considerable responsibilities on the customer. Other approaches to construction management are used by customers who want the advantages it provides but are unwilling to become closely involved in their projects. They use management contracting or possibly an approach called design and manage.

Management contracting is superficially very similar to construction management. The customer appoints a design team and a management contractor. They work with the customer to produce a brief, design and plan. These provide the basis for selecting the companies to undertake manufacturing, production and commissioning. The approach is different from construction management because the specialist contractors are employed by the management contractor as subcontractors rather than by the customer as works contractors. The cost to the customer is still determined by the cost of the specialist contractors' work. The difference is the customer pays these costs to the management contractor who passes the money on to the individual subcontractors. This saves the customer a great deal of administrative work in forming contracts and making many payments. It also gives the management contractor a contractual hold over the specialist contractors whereas construction management treats all the members of the project organization equally as the companies all have contracts directly with the customer.

Research into management contracting usually concludes the different status given to specialist contractors is inhibiting. They are seen to be less willing to focus on the project objectives or contribute ideas to the design or plan. Performance tends to be closer to traditional construction. There are exceptions and leading management contractors regularly match the performance of construction management but in general management contracting provides a less effective way of bringing management into construction projects.

There is a growing use of a variant of management contracting which somewhat confusingly is called construction management at risk. In this approach a construction management at risk company is selected when sufficient design information has been produced to allow a guaranteed maximum price for the manufacturing, production and commissioning to be established. The construction management at risk company ensures the design is completed in a manner which allows the new facility to be produced within the guaranteed maximum price. They then employ subcontractors in cooperation with the customer and design team, and manage the manufacturing, production and commissioning. The customer pays all the construction management at risk company's costs up to the limit set by the guaranteed maximum price. This approach limits the customer's cost risks although there are suggestions that construction management at risk companies have an incentive to inflate the guaranteed maximum price to reduce their own risks. The principle performance advantage is new facilities are produced faster than with traditional construction and in this respect it is similar to management contracting.

Design and manage is sometimes seen as a variant of design build rather than a management approach. In essence it means a management contractor takes responsibility for the design team as well as the specialist contractors. The approach may be led by design consultants but this is rare. The lead is normally taken by a management contractor. Its performance appears not to have been researched but is likely to be closer to that of design build than construction management.

7.10 The theory of construction management

This chapter provides a test of the theory of construction management. It is clear construction management provides opportunities for some of the propositions to be acted on and avoid some of the inherent weaknesses of traditional construction. The best available evidence suggests construction management achieves greater efficiency in terms of time than traditional construction. Since this is the customers' primary objective, it can be claimed construction management passes its second test and the construction management propositions in the section 'Construction management propositions' are supported. This is reflected in the example project which provides a further illustration of the use of the inherent difficulty indicators.

Example – Building Project using Construction Management

The example relates to a new hotel project similar to that used as the basis of the example in Chapter 5 but in this case using a construction management approach. The project starts with the experienced customer producing the brief and selecting the design and construction management teams. They work together in agreeing the design and plan and then selecting the works contractors. The resulting project team concentrates on reducing the production activities as much as possible. This is achieved by replacing production activities on-site with large prefabricated components manufactured off-site and simplifying design details to reduce the number of production teams needed. This strategy achieves shorter completion than the traditional approach and approximately the same as design build approach. It also results in a project which is inherently less difficult than similar projects using either of the other two approaches and this is reflected in the IDIs.

		Number of Teams	Number of team-days
Brief		1	60
Procurement	Select design and construction management teams	3	80
Design	Concept design	3	100
	Scheme design	6	150
	Detail design	8	180
	Technical design	8	200
	Production information	6	150
Plan		5	150
Procurement	Select companies	4	250
	Construction manager plan and control	4	250
	Contracts and payment systems	2	100
Manufacturing		10	250
Production	Substructure	3	100
	Structure	2	100
	External envelope	3	200
	Service cores	4	120
	Risers and main plant	4	100
	Entrance and vertical circulation	8	300
	Internal divisions	2	120
Decoration		5	500
Fittings		4	300
Landscaping and external services		3	250
Designer and construction manager's supervision		2	250
Commissioning		4	140
TOTALS		104	4,400

From equation 4.11 there are R = 5,356 possible relationships between teams working on this project. The detailed project plan, which can be seen on the website linked to this book, www.wiley.com/go/constructionmanagement-strategies, shows 3,230 of the possible relationships do not occur. This gives a total of R-k = 2,126 relationships that do in fact occur.

The largest number of teams involved in the same time interval is 20 so there are 190 different relationships.

The largest number of teams that could be involved in the same time interval is 34 in interval 21 so there could be 378 different relationships within that time interval. However, since not all teams work at the same time it is imperative to determine the number of teams that work simultaneously for each time interval. The plan shows that during the 21st time interval the following teams work simultaneously:

- 3 out of 4 construction manager's plan and control teams;
- 1 out of 2 contracts and payment systems teams;
- 4 out of 10 manufacturing teams;
- 3 out of 4 service core teams;
- 3 out of 4 risers and main plant teams;
- 5 out of 8 entrance and vertical circulation teams; and
- 1 out of 2 designer and construction manager's supervision teams.

This gives a total of 20 teams with 190 relationships.

A total of 90 teams have established relationships with some of the other teams and there are a total of 1687 established relationships. This gives an Established Relationships Indicator: $E_R = 1687/2126 = 0.79$ (equation 4.13). This is a good value which is then further confirmed by equation 4.14 which generates time dependent indicators of established relationships ranging from 0.51 to 1.00.

The relationship fluctuation indicator F_E is calculated using equation 4.15 and gives a value of 0.81. This reflects a strong influence of established relationships throughout the project because both E_R and F_E are comparatively high.

However, to really establish the benefits of established relationships we have to look at how long the teams involved are working together on this project and how much time they have spent working together on past projects. Equation 4.16 gives a Relationship Quality Indicator for each established relationship which ranges from 0.40 to 0.98 and the total Relationship Quality Indicator (equation 4.17) is: $Q_R = 1195/2126 = 0.56$. Although there are a number of high values for individual Relationship Quality Indicators, the total Relationship Quality Indicator is no more than reasonably good because there are 439 boundary relationships in this project out of possible 2,126.

From equation 4.18 we can calculate the relationships configuration complexity indicator C_R:

$$C_R = \frac{1}{3}\left(\left(1 - \frac{31}{207}\right) + \left(1 - \frac{64.2}{1990}\right) + \left(1 - \frac{20}{104}\right)\right) = 0.875$$

The past performance of the 104 teams over their immediately previous 10 projects provides individual team Performance Variability Indicators which range from 0.45 to 1.0. These give a total Performance Variability Indicator (equation 4.20): $R_p = 0.85$.

This is of course the average performance variability showing the teams managed to complete their work within the agreed time on 85 per cent of their recent projects.

The External Interference Indicator based on historical records for the particular region for the same type of building project undertaken by experienced project organizations using the construction management approach is 0.52.

Overall, we have the following inherent difficulty indicators which are compared to the indicators for the similar project using traditional construction:

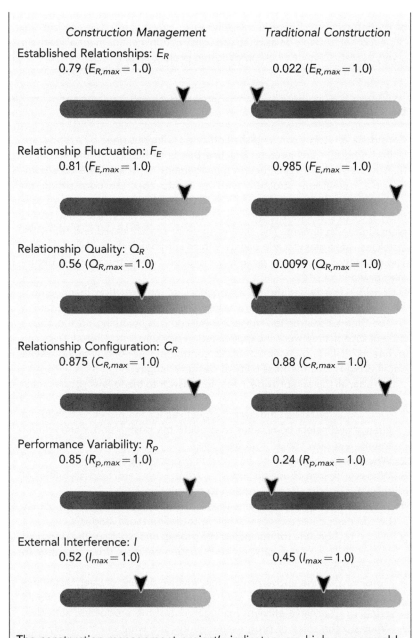

Construction Management	Traditional Construction
Established Relationships: E_R	
0.79 ($E_{R,max}=1.0$)	0.022 ($E_{R,max}=1.0$)
Relationship Fluctuation: F_E	
0.81 ($F_{E,max}=1.0$)	0.985 ($F_{E,max}=1.0$)
Relationship Quality: Q_R	
0.56 ($Q_{R,max}=1.0$)	0.0099 ($Q_{R,max}=1.0$)
Relationship Configuration: C_R	
0.875 ($C_{R,max}=1.0$)	0.88 ($C_{R,max}=1.0$)
Performance Variability: R_p	
0.85 ($R_{p,max}=1.0$)	0.24 ($R_{p,max}=1.0$)
External Interference: I	
0.52 ($I_{max}=1.0$)	0.45 ($I_{max}=1.0$)

The construction management project's indicators are high or reasonably high for Established Relationships, Relationship Quality, Performance Variability and External Interference. These indicators suggest the project has a manageable level of inherent difficultly. This is reinforced by the reasonably high values of the indicators for Relationship Fluctuation and Configuration which show good quality relationships in an appropriately optimised configuration tend to benefit most of the project. There is therefore a high probability the project will be completed much as planned.

This reflects the teams' reasonably good performance resulting from 90 teams with established relationships which benefit a significant proportion of the interactions. There is a likelihood of external interference but the project organization's decision to use construction management gives them a very realistic chance of completing the project as planned.

Exercise

Consider a 20-storey core and shell office building designed by a renowned United Kingdom architectural practice that is to be built in the centre of Manchester using a construction management approach. The core comprises a steel-frame structure with an in-situ concrete core sitting on prefabricated concrete piles and two levels of reinforced concrete basement, 6 elevator shafts and two sets of staircases, services, risers, and HVAC plants in the basement. The shell is made of prefabricated and ready-to-install insulated glass facade panels with an automatic internal blind system. The developer anticipates project completion 18 months after the beginning of work on site.

Consider the two differing approaches to construction management as depicted in Figure 7.2 and think what kind of difficulties could a United States construction management company working on the above project face if they were inexperienced in working with United Kingdom architects? [HINT: think about works contractors working with innovative detail designs and the possibility of design changes]

Consider all this and provide a detailed answer to the following questions:

1. Who should establish the overall objectives for the project, arrange the finance and select companies to provide the design and management teams, and how should that be done?
2. Why should customers be involved in project meetings?
3. Who is in charge of design and what are the roles and responsibilities of the design team (from inception to completion)?
4. Why is there a need for clear stages in the design process and what may be the main consequences of failing to define those stages?
5. Who is responsible for managing the project and what are the roles and responsibilities of the construction management team (from inception to completion)?
6. Consider that the above project may involve dozens of specialist works contractors and think what kind of a project organization you would establish to achieve efficient management of the production activities.
7. What are the roles of progress meetings and what are the likely consequences if agreed actions are not executed?
8. Describe the construction management process for the project taking into account that various different specialist companies may be involved at different stages to support customer decisions, and dozens of different works contractors may be involved in manufacturing and production.

Consider the following problem that may arise during the project and detail how a construction management company may respond in such a situation. Think why a situation like this can occur in the first place and what could be done to avoid similar changes in the future?

A considerable section of ventilation ducts have already been installed when a specialist designer announces that due to changed requirements in the upper 5 floors of the building a different ventilation system needs to be installed requiring wider air ducts.

Further Reading

The following publications are the source of ideas used in this chapter and provide further information for readers.

Construction Management Forum (1991) *Construction Management Forum: Report and Guidance*. Centre for Strategic Studies in Construction, Reading. This report was funded by leading construction management companies and describes best practice at a time when the approach was established as the most effective way of undertaking large, commercial developments.

Department of Construction Management (1979) *United Kingdom and United States Construction Industries: A Comparison of Design and Contract Procedures*. Royal Institution of Chartered Surveyors, London. This early report is based on research into construction management practice in the United States and was influential in persuading major customers and construction firms in the United Kingdom to adopt the approach.

Donohoe, S. and Brooks, L. (2007) Reflections on construction management procurement following Great Eastern Hotel Company v. John Laing. *Construction Management and Economics*, 25(7), 701–8. This paper describes an important legal case which helped clarify roles and responsibilities in the construction management approach.

El-Sayegh, S. M. (2009) Multi-criteria decision support model for selecting the appropriate construction management at risk firm. *Construction Management and Economics*, 27(4), 385–98. This paper describes the variant of the construction management approach which is called construction management at risk.

Chapter Eight
Partnering

8.1 Introduction

Partnering was developed in the West as it became apparent through the 1980s and early 1990s that Japan outperformed the West in key manufacturing industries, notably car manufacturing. As this superior performance was researched it became evident that in contrast to the West's reliance on competition and formal contracts, Japanese efficiency is based on cooperative, long-term relationships. Various initiatives designed to encourage cooperative teamwork in Western manufacturing were collectively called partnering.

The research into Japanese manufacturing was followed by similar studies of construction. These revealed Japanese construction companies had adopted the approach which gave Japanese manufacturing its international advantage and most importantly showed these highly impressive companies achieved reliably high levels of efficiency in terms of quality, time and cost. As a result partnering was introduced into construction, first in the United States. This innovation was quickly adopted by leading customers and construction companies in the United Kingdom where this distinctive approach developed rapidly. Partnering is now widely used in many countries.

Partnering is different from traditional construction, design build, management contracting and construction management in that it does not concentrate on defining effective roles and responsibilities. Instead partnering concentrates on establishing effective relationships in construction projects. It is based on the belief that the highest levels of efficiency result from everyone involved in a project using all the available time and resources cooperating to achieve agreed objectives.

Partnering thus rejects the common Western view that the highest levels of efficiency result from assembling project teams by inviting competitive bids for each individual activity, employing the specialist construction companies who submit the lowest bids, tying them into detailed contracts, and then insisting everyone works to the terms and conditions of those contracts.

Japanese construction relies on customers having long-term relationships with design build contractors who have long-term relationships with specialist subcontractors. This allows customers to state their requirements for a new facility, including the required completion date and total price in the confidence that everything will be delivered to high standards. The customer's established design build contractor begins productive work immediately. First, they search for the best possible way of meeting the customer's

Construction Management Strategies: A Theory of Construction Management,
First Edition. Milan Radosavljevic and John Bennett.
© 2012 John Wiley & Sons, Ltd. Published 2012 by John Wiley & Sons, Ltd.

requirements. Once the scheme design is agreed, the contractor produces a very detailed and complete design and plan which specifies what each specialist subcontractor will manufacture, produce and commission. The established specialist contractors are told exactly what they are to do in complete detail. This includes stating what work will be carried out each day on site. The design and plan are implemented relentlessly and the quality is checked continuously to ensure projects finish exactly on time and are absolutely complete when they are handed over to the customer. Major international comparisons of construction costs consistently show Japanese construction is the most efficient in the world.

Attempts to introduce the Japanese approach in the West inevitably hit many barriers formed by vested interests, established rules and regulations and firmly engrained opinions about the best ways of organizing construction. Partnering provides the most effective first steps towards matching Japanese levels of efficiency. It allows Western customers and construction companies to begin adopting better ways of working starting from their normal approach or from any other base.

As the initial steps are adopted and found to be effective, partnering provides further, more radical changes. This systematic, step-by-step approach provides at least three distinct levels of change which achieve ever higher levels of efficiency.

8.2 Project partnering

The first and most basic form of partnering provides a set of actions designed to be applied to an individual project by people new to partnering. It is called project partnering. The set of actions help customers and their project teams cooperate in finding ways of improving their performance for their mutual benefit.

A decision to use project partnering usually comes from discussions between an experienced customer and some of the construction companies they feel comfortable employing. The initial idea may come from the customer or one of the companies. The decision to use project partnering is made by the customer and the first step is to appoint all the key members of the project team early and ensure they are willing to work on the basis of cooperative teamwork. In practice this means they will take account of each others' interests in making decisions. The normal way of ensuring this is for the customer to hold a meeting with all the potential members of the initial project team to discuss how the project should be run. The aim is to reach a consensus on how they will work together. It is important everyone is open and honest about their own interests and concerns. This is often helped by inviting an expert in partnering to guide the meeting in establishing that everyone is genuinely prepared to work cooperatively.

First partnering workshop

Once a mature decision to use project partnering has been made and all the key members of the project team appointed, the first partnering workshop is held. This is the crucial device which enables partnering to be effective. It brings together all the key members of the project team, including the customer,

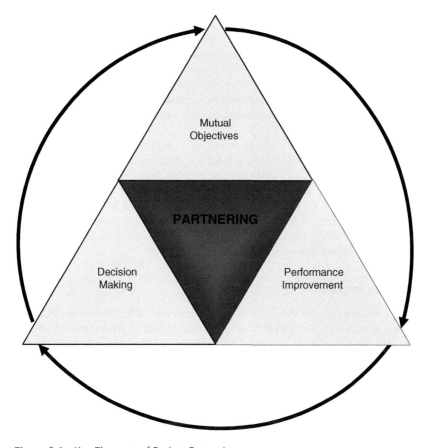

Figure 8.1   Key Elements of Project Partnering.

usually for two days. A professional facilitator guides the workshop to ensure the project's objectives and ways of working are fully and openly considered in a spirit of cooperation. The three essential outcomes from the first partnering workshop are shown in Figure 8.1.

Mutual objectives

The first objective for the workshop is to agree mutual objectives. The facilitator often has to begin by persuading tough, experienced businessmen they can benefit by cooperating. This means demonstrating it is possible to create so called 'win-win' agreements. Practitioners have to be convinced that if one party to a contract gains the other need not necessarily lose out in some way. So the aim is to cooperate in searching for a set of agreed objectives which give everyone more than they would expect from a project managed in the normal manner.

The mutual objectives may deal with almost any product or process related issues but must take account of the interests of everyone involved in the project. It is particularly important to sort out the financial arrangements so everyone gets a fair return for their work in normal business terms. On some projects the objectives deal with characteristics of the new facility including such issues as architectural quality, efficiency in use, maintenance and running costs. In other situations the objectives may deal with the process, including such things as: fast construction, low costs, reliable quality, excellent site facilities, safe construction, effective communication systems, focussed meetings and handing over a new facility fully complete. The mutual objectives usually deal with outcomes for the construction companies, which may include guaranteed profits, prompt payments, investment in communication systems, training in new skills, stream-lined management control systems, or any other matters important to the companies.

Best practice project partnering produces an agreed basic set of objectives which everyone is confident can be achieved. Making them explicit gives the customer and their project team a clear focus as they undertake the project. It enables everyone to insist that further consideration is given to decisions which threaten any of the mutual objectives. It is normal to have the mutual objectives recorded in clear, simple statements which are signed by everyone at the first partnering workshop and copies of the agreement distributed and displayed widely wherever project work is underway.

Decision making

The second purpose of the first partnering workshop is to agree the decision-making system to be used by the customer and project organization. Figure 8.2 illustrates an overall framework for decision-making systems, which the workshop should keep in mind. The agreed system needs to take account of the customer's requirements so, for example, an original design will require more

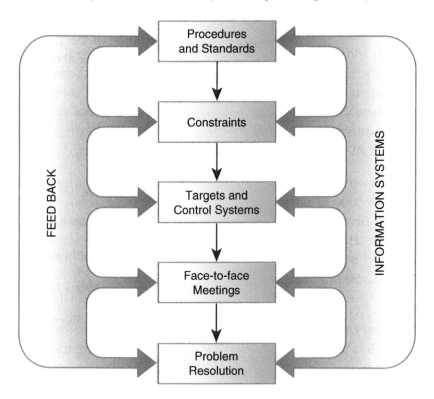

Figure 8.2 Project Decision-Making System.

discussion of a wider range of ideas than is needed where the customer decides to use an established design. The aims are to allow time for everyone who needs to be involved to: give full consideration to the key decisions; agree the way information will be communicated; decide the quality, time and cost control systems; ensure problems are dealt with quickly by those directly involved; and problems that persist are referred to more senior people to agree robust answers.

The processes need to ensure any failures are identified quickly. This is not to allocate blame, which is counter-productive. Failures should be used to guide the project team in agreeing how they can be avoided in future. An important part of dealing with failures is to agree how performance can be brought back to the agreed target levels as quickly as possible.

Success should also be identified and celebrated. When targets are being met, senior managers, including the customer, should congratulate and reward the teams involved. Nominating the best team performance each week and making a symbolic award can help ensure all teams strive to be winners.

It is important that ideas which result in improved performance are identified and recorded so they can be used on future projects. It is just as important to record the solutions to problems and failures so they can be avoided on future projects. All this feedback should be analysed and tested before being adopted in a different context. The overall aim is to guide the development of standards and procedures which reflect best practice.

Performance improvement

The third purpose of the first partnering workshop is to agree a specific, measurable improvement in performance and decide exactly how it will be achieved.

In the early stages of partnering there are two common mistakes. One is to aim at a modest improvement which, at the end of the project, provides a poor return for the effort involved in cooperating. The other common mistake is to aim at a raft of improvements. Any one of them could have been achieved but together the ambitious targets ask too much of a project team. Both mistakes lead to partnering being dismissed as ineffective.

It is important for project teams to concentrate on one significant improvement. It is even more important they do not allow this focus to damage their established normal performance in other areas. Thus, if the main objective is to reduce costs by 10 per cent, it is important not to let quality or safety or time control deteriorate. The task at the first partnering workshop is to agree one significant improvement in performance and ensure the controls needed to maintain normal performance are in place, fully understood and will be put into practice.

The agreed performance improvement should be included in the published list of mutual objectives as a constant reminder to everyone involved with the project.

Partnering workshops

Best practice holds partnering workshops throughout projects to review progress and consider ways of ensuring the project team achieves its mutual objectives and performance improvement. They are organized on the same basis as the first workshop and aim to reinforce cooperative behaviour.

Each workshop is guided by a robust progress report. This may show the project is going better than planned and the team may decide it can aim for bigger performance improvements. This needs to be discussed and a well-considered decision made. It may be sensible to stick to the original target and make sure it is met. On the other hand if the project organization can achieve more and the workshop agrees actions and a fair distribution of the resulting benefits, this will help build commitment to the project and partnering.

Equally there may be problems the team decides it must solve. This is not to usurp the agreed decision-making system but to identify and tackle persistent problems which are inhibiting cooperative behaviour. The workshop should continue until they have agreed a permanent solution because nothing undermines partnering as surely as an unresolved, persistent problem.

A final workshop is held to identify all the good ideas and lessons from the project and decide what should be done with each of them. Some will need further development and testing, some may be project specific and are not likely to be useful elsewhere, but some will be fit for wider use. These should be recorded and steps taken to make them available for future projects.

8.3 Strategic partnering

Partnering can be effective on individual projects. It does however involve considerable costs in selecting construction companies willing to cooperate, and running effective workshops. The efforts involved in ensuring the whole project team works together on the basis of cooperation may take time. Some of the selected construction companies may find it difficult to overcome their ingrained competitive instincts. A first experience of partnering often requires individuals to tackle a steep learning curve illustrated in Figure 8.3.

The benefits can be worthwhile and exceed the costs on an individual project but it is the case the benefits of partnering are easier to achieve on a second, third and subsequent projects. This is why many groups of companies having experienced successful project partnering decide to continue working together on a series of projects. This is usually called strategic partnering. The actions which distinguish strategic partnering are sometimes called the seven pillars of partnering because they comprise the seven sets of partnering actions shown in Figure 8.4.

Strategy

Strategic partnering is guided by an explicit strategy agreed by the cooperating companies. It establishes the kind of buildings or infrastructure to be constructed. This is usually self-evident because the companies came together to undertake a series of projects providing a particular kind of new facility. Nevertheless it is important that each company considers its strengths and weaknesses in respect of the particular facility and for them to have a joint discussion of their expectations. The companies need a common understanding of the facilities they will construct. They should agree the standard of their products, the technologies they will use, how the facilities will be used, the life-cycle costs and the environmental impacts.

The companies should discuss and agree the services they intend providing in support of their products. These may include setting up interactive

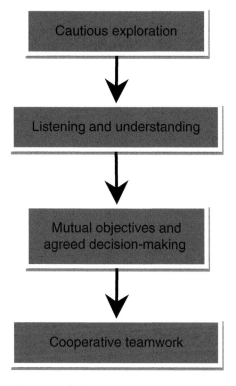

Figure 8.3   The Development of Effective Teams.

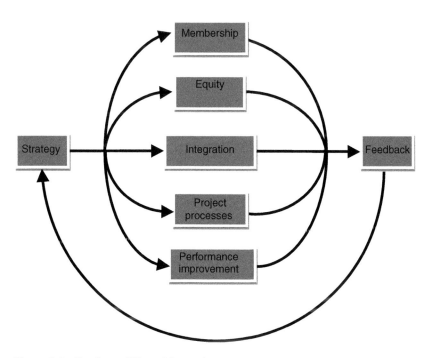

Figure 8.4   The Seven Pillars of Partnering.

demonstrations of possible new facilities to involve the intended users in discussions of the brief and design. They may include access to finance, site visits for users, training in running the new facility, and maintenance services.

The strategic decisions need to be based on sound business principles. There should be clear plans for performance improvements which translate into lower costs or higher value and so generate bigger profits. There should be a firmly-based agreement on how benefits will be distributed among the participating companies. Similarly the companies need to agree how financial problems and losses will be tackled.

In reaching their strategic agreement, the companies must ensure it fits their individual organizations' capabilities. The strategy should be sufficiently flexible to cope with changed circumstances and to encourage development and innovation.

Membership

The companies in an initial strategic partnering arrangement are chosen because they want to work together. Even where companies have experience of working together, it is good practice to evaluate each of them before setting up a strategic partnering arrangement. The evaluation needs to consider technological skills and knowledge, business characteristics, and cooperative attitudes. Partner companies' strengths should compliment rather than duplicate each other so the selected companies can cooperate to form a fully competent organization. They need to be able to put the strategy into effect, have the potential to continually improve their joint performance, and relentlessly develop new ideas.

The evaluations should be repeated at regular intervals to ensure all the companies are contributing to the strategic partnering arrangement's success. It is often the case that changes to the product or market lead some of the original companies to decide the arrangement no longer benefits them or it may become apparent that one or more of the original companies is no longer contributing effectively. Whatever the causes, if the need for change becomes apparent, it must be discussed openly, a decision made and action taken. Any partner company required to leave should be treated generously. Their real concerns and problems should be dealt with sympathetically. Fair treatment is important in maintaining the commitment of the remaining companies and it helps in maintaining a good reputation with customers, suppliers and potential new partner companies.

Equity

The financial arrangements should encourage the partnering companies to invest in long-term development aimed at improving their joint performance. This means the companies get more inside the strategic partnering arrangement than they could reasonably expect elsewhere. It is unlikely that this means the benefits are shared equally because normal profit levels for consultants, contractors and specialists vary considerably. A more difficult issue is the customer's benefits derive from the impact the new facility has on their business and this may be impossible to predetermine. Taking all the practical issues into account, the aim should be for the shares to be regarded as fair by all the companies. This issue deserves to be discussed very carefully because the financial arrangements for the strategic partnering arrangement need to be robust enough to withstand the inevitable problems and crises. Once agreed the financial arrangements should form the basis for an explicit business plan endorsed by all the companies.

As the arrangement develops and takes new initiatives, the financial arrangements often have a direct influence on decisions. Individual companies will have access to different sources of finance subject to very different terms and conditions. Banks and shareholders may view partnering very differently. Some will recognise the benefits but others may regard it as risky. They may question the need for cooperative attitudes and insist the companies must seize every opportunity to earn the biggest possible short-term profits. It is important that these issues are discussed inside the strategic partnering arrangement and decisions made in the joint best interests of all the partnering companies. In these discussions, companies may need to be reminded that partnering depends on seeing problems as opportunities to identify improvements and basing decisions on the long-term interests of the companies. These attitudes are essential for the long-term strength of strategic partnering arrangements.

Integration

It is important to agree how all levels of the partnering companies will work together. This takes time and experience of working in cooperation on a number of projects. The driving force for the strategic partnering arrangement is provided by a strategic team of senior managers from all the companies. The strategic team provides leadership and makes strategic decisions which go beyond individual projects.

Mature strategic partnering organizations behave like a single organization. All construction actions use well-integrated systems based on common information and communication technologies. Meetings have clearly defined purposes and people attending know the background and are properly briefed on the key decisions to be discussed and decided. Social events build team spirit and give people a rounded picture of colleagues. Team offices are used to bring together everyone involved in a distinct stage of a project, irrespective of which company employs them. All this serves to encourage open and effective decision making.

Some of the most effective strategic partnering arrangements establish interface teams comprising the managers in each of the companies responsible for a specific activity. The interface teams meet regularly to deal with problems, identify and introduce improvements to their activities, and consider innovations capable of making strategic change to the companies' joint performance, products or services. Thus, interface teams may deal with design, technology, quality, time, cost, safety and other issues. Finance is nearly always important and it is normal for an integration team to bring together the chief financial managers from the separate companies. They may well identify major economies in the way financial matters are dealt with. For example, they may decide to accept project costs as calculated by the project team without the need for double checking or separate audits. Savings of this kind can provide large improvements in efficiency by, for example, halving the number of managers needed to run projects, providing more reliable information, and supporting good decision making.

Project processes

All the pillars aim to improve project processes so they deliver consistently high levels of performance. This pillar serves to emphasise the central importance of undertaking projects efficiently.

The broad aim is for project organizations to be supported by standardised actions, processes and technologies which represent current best practice. This helps ensure project organizations can be assembled quickly and then well-drilled construction teams can carry out their work virtually automatically.

Achieving high efficiency requires standards to be applied consistently. It also requires standards to be reviewed and improved on the basis of carefully tested improvements. This development work should be undertaken outside any individual project in order to achieve the highest levels of efficiency. Ideas for improvements often emerge from individual projects but they may equally come from many other sources. All potentially promising ideas should be thoroughly developed and tested on one or more projects in carefully controlled trials before being incorporated in established standards.

Projects which require original designs or use innovative technologies pose particular challenges for standardised project processes. It is nevertheless important for creative activities to take place within a controlled framework. This needs to be flexible to allow for difficult problems but construction projects should be completed when the customer needs the new facility at a cost they can afford and so there have to be limits to the time and resources devoted to finding better answers. The best strategic partnering arrangements devise well-thought-out standards which allow the initial stages to be free-ranging and creative but then provide a secure basis for controlled work.

Performance improvement

The whole point of partnering is to relentlessly improve performance. It is therefore important for the strategic partnering organization to ensure it is making continuous, measurable improvements, particularly in terms of the agreed strategic objectives.

It helps to use benchmarks based on established industry norms to measure performance. This allows the organization's joint performance to be compared with current best practice. It also provides an important balance which ensures the internal cooperative culture does not lead to the complacency and inefficiency which can creep into non-competitive environments.

The benchmarks should measure all the key aspects of performance not just those where improvements are being pursued. This ensures that no parts of the joint organization focus too much on improvements at the expense of the more routine aspects of overall performance. Strategic partnering organizations commonly measure: customer's satisfaction with their new facilities and the services they have received; the number of recorded defects at handover; quality; time; costs; the achievement of budgets and completion dates; safety; productivity; and profitability.

Benchmarks measure performance and where the actions described in the other pillars are put into effect, performance steadily improves. Occasions arise when a major step forward in performance is needed. An effective approach in these circumstances is for the strategic team to set up a task force. This is a group of specialists brought together to solve a problem, provide the basis for a process improvement, develop a new design approach, or take some other initiative intended to deliver a significant performance improvement.

Feedback

Feedback turns networks into controlled systems. The companies involved in a strategic partnering organization form networks of teams at various levels of responsibility to undertake all the various stages of projects and deal with strategic issues. These networks all need feedback to guide their decisions.

Feedback should tell the strategic team whether their strategic objectives are being met. Individual companies need feedback to help them assess the continued benefits of the partnering arrangements. Project teams depend on feedback to ensure they are achieving their agreed objectives and to highlight potential problems. Construction teams need feedback on their performance against agreed targets. All construction actions should be guided by feedback which identifies problems early, measures progress and relates performance to agreed objectives. These various types of feedback require systems which collect the relevant data, analyse it and deliver it to the individuals who need it in time for them to use the information effectively.

The formal systems should be supplemented by senior managers, especially members of the strategic team, regularly visiting offices, factories and construction sites where the joint organization's work is underway. They should ask questions of the people doing the work and be prepared to answer questions about the strategic partnering arrangement. The resulting first-hand knowledge should be used to interpret and if necessary question the formal feedback.

The strategic team should have feedback from the facilities they construct. This should capture the views and experience of customers, users, owners and the people who run and maintain the buildings or infrastructure.

Feedback provides the basis for reviewing the overall strategy, setting new targets, identifying persistent problems, guiding innovations and new developments, and ensuring future actions deliver performance improvements. It guides project organizations in achieving ever higher levels of efficiency. It helps individual construction teams identify weaknesses and ways of improving their own knowledge, skills and performance.

Internal partnering

As a strategic partnering arrangement becomes established, the companies involved normally find they need to make changes which support partnering. This is often called internal partnering which is most effective when an internal partnering team of senior managers is established to lead the company's use of partnering. Its role is: to establish a genuine commitment to partnering from top management; support senior managers as they consider the use of partnering on new projects; and ensure their construction teams are supported in using cooperative teamwork.

Achieving these aims may require changes to the company's established ways of working. Controls driven by the dictates of market competition are likely to obstruct cooperation. New ways of judging costs and benefits may be needed to reflect the benefits partnering can deliver over a series of projects. It is sensible to undertake a review of the company's strengths and weaknesses to give a robust basis for making commitments at partnering workshops. This should particularly evaluate the company's personnel and their success in working together and with other companies. The local construction market should be studied to establish the likely level and nature of demand for the types of projects which best suit the company. It is also worth researching the attitude of customers and other construction companies to partnering. All of this should be brought together in a detailed business plan produced by the internal partnering team.

A key part of the business plan should provide mechanisms which help the internal partnering team establish partnering throughout the company. This is not to suggest companies suddenly abandon their established markets and ways of working because that could well be an unacceptable risk in many situations. The aim is to ensure departments and divisions within the company do not obstruct partnering. Established rules, attitudes and methods may create unnecessary barriers which handicap teams involved in setting up new partnering projects. The internal partnering team of senior managers need to discuss anything which obstructs partnering and if at all possible ensure all parts of the company support the new approach.

As internal changes are introduced the internal team of senior managers need to consider their impact on the other companies in their existing partnering arrangements. Unilateral changes can easily create unexpected problems for partners. It is best to actively ensure changes are discussed, agreed and coordinated with partners. Working together in this way can provide unexpected benefits and new ideas as a group of companies search for the best ways of working together.

The internal team of senior managers should review existing partnering arrangements to ensure they remain effective and are delivering real benefits. Like many human organizations, partnering arrangements have life cycles. They emerge, develop, reach maturity, face crises and change or decline. The internal team need a clear view of the stage reached by each of the company's partnering arrangements. This should take account of the formal feedback from the partnering teams supplemented by walking around the places where partnering projects are underway and asking questions aimed at determining how well individuals understand the mutual objectives, decision-making processes and agreed performance improvement.

At times the ongoing review of partnering arrangements will identify a need for radical changes. The company may need to: concentrate on a different type of construction project; change its internal structure or even its senior managers; give more emphasise to external rather than internal communications; become more flexible to allow a wider range of projects to be undertaken effectively; invest in education and training; develop close ties with local communities; or withdraw from some existing partnering arrangements.

Individual training is often an important part of establishing partnering throughout a company. This may include developing communication skills, practicing cooperative decision-making and updating technical knowledge to ensure construction teams are competent. It is sensible to provide training to complete construction teams so they progress together as an integrated unit. The training should be reinforced by teams being given greater authority to make decisions and enter into commitments in step with their improving knowledge and skills. Achieving these major changes may well require teams to be given specialised training in, for example, the use of quality, safety, time and cost control systems.

In all of their actions the internal team of senior managers should balance the company's long-term stability with the flexibility needed to respond to new opportunities and challenges. An important part of this is balancing the advantages of specialised head-office departments with the benefits of encouraging independent construction teams. Similarly the benefits of using standard answers need to be balanced with the long-term need for original answers and innovation. Both are important in the right situations but they require very different construction teams and project organizations. Working out effective answers to these difficult strategic issues leads some companies to restrict their use of partnering to situations where it provides robust benefits. Other companies embrace partnering fully and move on to what is sometimes called third generation partnering but more usually is known as strategic collaborative working.

8.4 Strategic collaborative working

Strategic collaborative working means a group of construction companies cooperating to develop a long-term business. The companies may come together because they have recognised an attractive business opportunity. Perhaps more commonly strategic collaborative working emerges from a group of companies which have used strategic partnering successfully over a number of projects. Having experienced the benefits of cooperating, they decide there are further benefits in a more permanent collaboration.

The companies have formal agreements which give them the confidence to invest in developing a product and support services. The product is a specific type of building or infrastructure facility. Typical examples include apartments, housing, student residences, warehouses, supermarkets, fast-food outlets, schools, hospitals, sports stadia and motorways. These are often marketed under a brand name backed by a range of customer support services.

The formal agreement is usually set out in a governing document which establishes the principles that dictate how the companies will deal with each other. It embodies the decisions made in respect of each of the seven pillars of partnering. As with strategic partnering, strategic collaborative working is led by a strategic team which ensures the principles are acted on in all the companies. The senior managers who form the strategic team work through an organization formed by teams drawn from the participating companies as illustrated in Figure 8.5.

The success of strategic collaborative working is closely related to the joint organizations' ability to set and meet tough targets. Cases exist where remarkable progress has been achieved in response to challenging targets. These include such things as: halving the cost of a supermarket while maintaining an image of quality; reducing site labour by 80 per cent; eliminating all site accidents; achieving zero defects at every stage of construction; reducing the customer's labour force needed to run a new facility by 25 per cent; and constructing a fast food outlet on a prepared site in 24 hours.

Where companies have the confidence to give a joint team the time and resources to tackle challenging targets, wonderful things can happen. Technologies are merged. Prefabrication is used to reduce site activities to a rapid

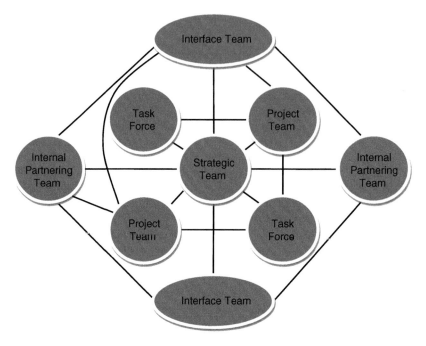

Figure 8.5 A Strategic Collaborative Working Organisation

and highly productive set of activities. Standard components replace individual design and manufacture. Mechanical systems are replaced by electronic devices. Design and manufacturing are merged. Construction becomes a modern industry delivering reliable quality and high levels of efficiency.

Success brings its own challenges. The ownership of new technologies and services can be controversial. The most effective approach is to allow all the partnering firms to use innovations inside or outside the strategic collaborative working arrangement. Provided one company's use of an innovation outside the joint organization does not inhibit its use inside, this is allowed. The key to this working well is the companies being open with each other so there are no surprises.

Other problems arise when agreed actions suit the internal organization of some companies better than others. This challenges individual companies to tolerate some ambiguities and contradictions. They may have to accept different ways of working in different parts of the company. The resulting flexibility can be a decisive strength when market conditions change because companies that accept a variety of behaviour around a central core of focussed work tend to survive better than competitors who concentrate on single-minded super-efficiency.

Strategic collaborative working which produces brand named products and support services naturally leads construction companies into investing in modern marketing. They want to attract demanding customers to buy their products and services. Marketing does not require huge advertising campaigns or an expensive sales force. The aim of marketing should be providing reliable information to potential customers. It means working with customers to establish their real needs, making the choices easy to understand and delivering on every explicit and implicit promise. Marketing can be seen as the final piece of the jigsaw which turns construction into a modern consumer product industry.

8.5 Partnering efficiency

Partnering aims to provide improvements to efficiency. Unlike design build and management approaches, partnering does not predetermine any particular pattern of organization. It could therefore be argued it does not provide a basis for testing the theory of construction management. However, partnering does result in deliberate changes. Its purpose is to provide sets of actions that encourage companies, project organizations and construction teams to agree ways of improving their efficiency. The outcomes are shaped by the pre-existing knowledge and skills of the companies, project teams and construction teams involved. This is more than sufficient for the chosen actions and their outcomes to provide a wide ranging test of the theory of construction management.

It is therefore fortunate that case studies of effective partnering describe many initiatives and the resulting impacts on efficiency. The following examples taken from case studies reported in the Further Reading listed at the end of this chapter illustrate the kind of actions taken when partnering is well used.

Strategic partnering to construct a series of supermarkets used value-engineered design details to reduce the number of individual construction teams involved in the projects and minimise the interactions between them.

Another supermarket chain took a different approach to reducing the number of individual construction teams and minimising the interactions between them. They used partnering workshops to review their established design approach and identify ways of combining or eliminating technologies.

A strategic partnering arrangement to build high-quality warehouses quickly and at lower costs than their competitors developed a set of principles which ensured everyone involved made a substantial contribution to meeting the objectives and reducing the need for interactions between them. The principles included: base the brief for each new project on a previous design that had worked well; appoint the whole project organization early; use standard design details; divide the facility into zones which fit seamlessly together so each can be designed, planned and produced independently by specialists; have pre-determined agendas for standard meetings; give control of time and quality the highest priority; and use large-scale prefabricated components and elements wherever possible.

A major oil company had used what was in effect a partnering approach with serial contractors to halve the time taken to produce filling stations using traditional site based technologies. A key factor in this achievement was using the same experienced production teams on every project. A point was reached when no further improvements in performance were possible using the traditional technologies. The oil company then used strategic partnering with two carefully selected companies to standardise and prefabricate the shop element of their filling stations across Europe. This removed one of the most complex elements of their projects which reduced the number of individual construction teams involved and minimised the interactions between them.

Specialist subcontractors were involved in the design of a project where the customer wanted a new training facility to include innovative ways of reducing energy use and minimising the environmental impact. Partnering enabled this innovative facility to be constructed close to the predicted cost, more quickly than expected largely because components fitted together easily with minimal material waste. This reduced the variability delivered by the construction teams involved.

A controversial road building project was completed early and below budget by including local companies and residents in the open communication which was a key feature of the project organization's use of partnering. This insulated the project from interference which on a traditional project may well have resulted in substantial delays and claims for additional costs from the construction companies involved of the order of 50 per cent of the total cost.

A housing association decided to partner with a contractor because they had experience of a steel frame system which provided substantial reductions in production times. The contractor had already established a good track record of working with the housing association and this gave both parties confidence in using the innovative technology. Realising that innovation inevitably throws up problems, the partners decided it was necessary to improve the quality of support provided to the construction teams involved. They did this by having each company appoint a senior manager to expedite decision making and ensure their construction teams were given all the support they needed to meet the project objectives.

A partnering project to construct a new police station involved suppliers and subcontractors early so their advice on the most effective use of their technologies could be incorporated in the design. This produced an amazing number of innovations and ensured the construction teams working on site fully understood the technologies used.

The agency responsible for maintaining and operating a highway network decided they needed to improve the performance of their construction projects. They approached this by selecting consultants and contractors able to provide construction teams competent in all the technologies required by their projects. A series of partnering workshops produced ideas from unexpected sources within the selected companies and established radically new ways of working which benefited road users and ensured public funds were spent wisely.

A major bank used partnering to assemble seven project teams to construct regional centres providing specialist banking services. They did this by inviting medium-sized construction companies they had previously worked with on similar projects to bid for the work. They then selected specific individuals they trusted from within the successful companies to form the core team of each project organization. The whole programme was completed with no contractual disputes, on time and within budget.

A water company organized partnering workshops for all the consultants and contractors involved in the refurbishment of a treatment plant. A particularly successful initiative was including key decision makers from head office and the staff who operated the plant in the workshops. As a result the construction teams and the operatives who ran the plant were fully involved in agreeing the project's objectives and were motivated to achieve them.

A strategic partnering arrangement to produce runways and aprons at the United Kingdom's leading airports developed standard detail design drawings and specifications, and generic construction plans which helped ensure accurate communication between the teams. They ensured these were well used by setting up a common office for the construction teams from all the partnering companies. This had many benefits and in particular minimised the effort required to achieve accurate communication between the teams.

A major university used partnering with companies installing energy saving systems in busy academic buildings where continuity of energy supplies was vital. A key to the successful completion of this difficult project was the early attention given to establishing a framework for accurate communications between the various companies involved in the project and between the users and the contractors working in their buildings. This framework included a simple mechanism for holding a meeting as soon as any problem arose and a policy which stipulated that all problems had to be solved straight away.

A company responsible for a national chain of hotels and restaurants used strategic partnering with six regional contractors and their specialist subcontractors. Regular workshops brought all the partners together to review their methods and study particularly successful projects. This led to the development of standardised and simplified technologies, and a genuine team-spirit. One significant consequence was a dramatic reduction in the amount of communication needed between the construction teams undertaking each project.

A customer and a design build contractor established a secure basis for their project partnering by using a standard design build contract well understood by both parties. They kept key terms simple by agreeing a target cost with savings or cost over-runs split 50:50 between the customer and contractor. They also agreed an ambitious completion date but provided for only nominal liquidated damages of £1 a day to be paid by the contractor if the date was missed. Their deliberate aims were to: minimise the length and intensity of negotiations needed to agree the contract; ensure both parties regarded the outcomes as advantageous to themselves; and ensure there was no temptation for anyone to waste time and resources trying to improve the terms of the agreement.

8.6 Construction management propositions

These various actions taken from case studies of successful partnering provide support for the following construction management propositions.

To enable construction to be undertaken efficiently within a partnering approach, construction management aims to:

- Reduce the number of individual construction teams involved in a construction project.
- Improve the quality of relationships between the construction teams involved in a construction project.
- Reduce the performance variability of the construction teams involved in a construction project.
- Reduce the external interference experienced by the construction teams involved in a construction project.
- Select construction teams competent in the technologies required by the project in which they are involved.

- Ensure construction teams accept the agreed objectives of the project in which they are involved.
- Ensure construction teams are motivated to achieve the agreed objectives of the project in which they are involved.
- Foster accurate communication between the construction teams involved in a project organization.
- Minimise the effort required to achieve accurate communication between the construction teams involved in a project organization.
- Minimise the length and intensity of negotiations needed to agree the transactions which bring construction teams into a project organization.
- Ensure construction teams regard the transactions which brought them into a project organization as advantageous to themselves.
- Minimise the resources construction teams devote to improving the terms of the transactions which brought them into a project organization.

8.7 Partnering performance

There is a large body of research into successful partnering arrangements which describe project organizations relentlessly finding better ways of working. Table 8.1 shows the level of cost and time reductions which have been achieved on projects using partnering effectively. These results come from measuring the performance achieved on similar projects by similar or in many cases the same companies before and after partnering was adopted.

The table records the highest levels of efficiency reliably achieved by projects using partnering and reported in serious case studies. Overall the results from case studies show a range of improvements in costs and times. For example, project partnering provides cost reductions which range from 5 per cent to 30 per cent. The figures for strategic partnering are cost reductions of 10 per cent to 40 per cent; and for strategic collaborative working, 10 per cent to 50 per cent. The results for time reductions show less variation but are dependent on the specific objectives adopted by the companies involved. It is apparent that time reductions are easier to achieve than cost reductions and almost any level of improvement is possible given a committed group of companies using partnering.

The greater efficiency which enables partnering companies to reduce costs and times may be used to provide other benefits. Case studies of partnering projects report many such benefits including: significant increases in customer satisfaction; radical reductions in administrative work for customers; reliable completion on time; higher productivity at all stages of projects; much better safety records on construction sites; increased profits for

	Cost reduction (%)	Time reduction (%)
Project partnering	30	40
Strategic partnering	40	50
Strategic collaborative working	50	80

Table 8.1 Cost and Time Reductions Provided by Partnering

Sources: Bennett and Jayes (1998) and Bennett and Peace (2006)

construction companies; more predictable workload for construction companies; substantial reductions in the number of defects in completed facilities; avoiding claims for additional costs by construction companies; reductions in the amount of waste material produced by construction sites and much higher proportions being recycled; initiatives which effectively minimised the effects of exceptionally bad weather on production activities; reduced disruption for the users of existing facilities which were being altered or extended; and feedback which gave construction teams reliable measures of their own performance in time to understand and build on strengths and identify areas needing improvement.

Partnering provides this variety of benefits because all the parties with an interest in a project are encouraged to agree mutual objectives. Partnering then provides a means for the parties to work out the most efficient ways of achieving their objectives whatever they happen to be. The key to successful partnering is to work together in a spirit of cooperation. It is not enough for a customer and construction companies to say they are partnering. They have to cooperate in creating and sustaining a culture which relentlessly aims at higher levels of efficiency.

8.8 The theory of construction management

This chapter provides a further test of the theory of construction management. Partnering encourages project organizations to act on some of the propositions and avoid some of the inherent weaknesses of traditional construction. There are many case studies and careful research which provides evidence to suggest partnering enables project organizations to achieve greater efficiency than traditional construction. This provides strong support for the construction management propositions listed in the section 'Construction management propositions'.

Partnering was devised as a means of introducing distinctive features of the successful Japanese approach into Western construction industries. In other words it is no more than the first steps towards the highest levels of efficiency. Strategic collaborative working does enable construction companies to approach the levels of efficiency routinely achieved by the leading Japanese construction companies and indeed those in a few other countries which have adopted a similarly totally integrated approach. The most important Western example is Germany which like Japan has invested in highly efficient manufacturing. Construction has copied many of the key features. The resulting total construction service is described in Chapter 9.

Example – Building Project using Partnering

The example relates to new hotel project similar to that used as the basis of the example in Chapter 5. The project organization uses an essentially traditional approach but the companies responsible for the project decide to use a project partnering approach. This enables them to achieve significant economies throughout the project which are reflected in the following data and calculations. The overall effect is the partnering project is likely to be completed faster and cost less than the traditional construction approach.

		Number of Teams	Number of team-days
Brief		1	80
Procurement	Select design companies	1	40
Design	Concept design	1	100
	Scheme design	1	80
	Detail design	5	100
	Technical design	5	100
	Production information	2	80
Plan		2	100
Procurement	Select companies	3	100
	Plan and control	4	100
	Contracts and payment systems	1	30
Manufacturing		10	200
Production	Substructure	3	100
	Structure	2	100
	External envelope	3	200
	Service cores	4	120
	Risers and main plant	4	100
	Entrance and vertical circulation	8	300
	Internal divisions	2	120
Decoration		5	500
Fittings		4	300
Landscaping and external services		3	250
Designer and construction manager's supervision		2	160
Commissioning		4	140
TOTALS		80	3,500

According to equation 4.11 there are $R = 3,160$ possible relationships between teams working on this project. The detailed project plan, which can be seen on the website linked to this book, www.wiley.com/go/constructionmanagementstrategies, shows 2,038 of the possible relationships do not occur. This gives a total of $R-k = 1,122$ relationships that do in fact occur.

The largest number of teams that could be involved in the same time interval is 23 in interval 21 so there could be 253 different relationships. However, since not all teams work at the same time it is imperative to determine the number of teams that work simultaneously for each time interval. In this case the plan shows during the 21st time interval the largest number of teams work simultaneously:

- 2 out of 4 plan and control teams;
- 3 out of 4 service core teams;
- 4 out of 4 risers and main plant teams;
- 6 out of 8 entrance and vertical circulation teams;
- 1 out of 1 contracts and payment systems teams; and
- 2 out of 2 designer and construction manager's supervision teams.

This gives a total of 18 teams with 153 relationships.

Every team on a project has established relationships with at least one but often several other teams giving a total of 1,023 established relationships. This gives an Established Relationships Indicator: $E_R = 1023/1122 = 0.91$ (equation 4:13). This is a very good value which is then further confirmed by equation 4:14 which generates time dependent indicators of established relationships ranging from 0.55 to 1.00.

The relationship fluctuation indicator F_E is calculated using equation 4.15 and gives a value of 0.86. This reflects a strong influence of established relationships throughout the project because both E_R and F_E are comparatively high.

However, to really establish the benefits of established relationships we have to look at how long the teams involved are working together on this project and how much time they have spent working together on past projects. Then equation 4:16 gives a Relationship Quality Indicator for each established relationship which ranges from 0.70 to 0.98 and the total Relationship Quality Indicator (equation 4:17) is: $Q_R = 958/1122 = 0.85$. The high number of established relationships together with the number of high values for individual Relationship Quality Indicators means the total Relationship Quality Indicator is good because there are only 99 boundary relationships in this project out of possible 1,122.

According to equation 4.18 we can calculate the relationships configuration complexity indicator C_R:

$$C_R = \frac{1}{3}\left(\left(1 - \frac{31}{159}\right) + \left(1 - \frac{50}{1550}\right) + \left(1 - \frac{18}{80}\right)\right) = 0.85$$

The past performance of the 80 teams over their immediately previous 10 projects provides individual team Performance Variability Indicators which range from 0.65 to 1.0. These give a total Performance Variability Indicator (equation 4:20): $R_p = 0.89$.

This is of course the average performance variability showing the teams managed to complete their work within the agreed time on 89 per cent of their recent projects (note we have assumed c_j to be 1 for all teams although for real life projects these values should be determined separately).

The External Interference Indicator based on historical records for the particular region for the same type of building project undertaken by experienced project organizations using partnering is 0.74.

Overall, we have the following inherent difficulty indicators which are compared to the indicators for the similar project using traditional construction:

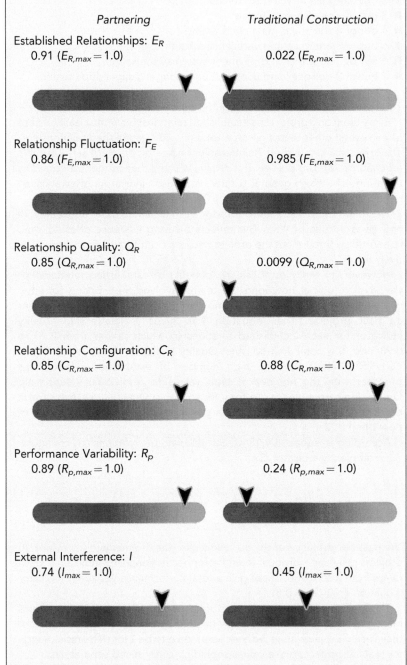

	Partnering	Traditional Construction

Established Relationships: E_R
0.91 ($E_{R,max}=1.0$)   0.022 ($E_{R,max}=1.0$)

Relationship Fluctuation: F_E
0.86 ($F_{E,max}=1.0$)   0.985 ($F_{E,max}=1.0$)

Relationship Quality: Q_R
0.85 ($Q_{R,max}=1.0$)   0.0099 ($Q_{R,max}=1.0$)

Relationship Configuration: C_R
0.85 ($C_{R,max}=1.0$)   0.88 ($C_{R,max}=1.0$)

Performance Variability: R_p
0.89 ($R_{p,max}=1.0$)   0.24 ($R_{p,max}=1.0$)

External Interference: I
0.74 ($I_{max}=1.0$)   0.45 ($I_{max}=1.0$)

The partnering project's indicators are high for Established Relationships, Relationship Quality, Performance Variability and External Interference. These indicators suggest the project has a low level of inherent difficulty.

This is reinforced by the high values of the indicators for Relationship Fluctuation and Configuration which show good quality of relationships in an appropriately optimised configuration tend to benefit most of the project. There is therefore a high probability the project will be completed as planned.

This reflects the teams' reasonably good performance resulting from a relatively large number of established relationships which benefit a significant proportion of the interactions. There is some likelihood of external interference but the project organization use of partnering gives them a very good chance of dealing with the resulting problems and completing the project as planned.

Exercise

Imagine you run a large development company intending to build a new 30-storey multi-purpose tower, which is to house an underground externally operated 3-storey car park, and then above ground a shopping zone on the first 5 floors, a hotel on the next 5 floors, while the rest of the building is dedicated office space. Take into account the development company has built and has plans in place to build many more similar projects all over the world. The building is a steel frame structure surrounding an in-situ concrete core sitting on prefabricated concrete piles and in-situ concrete basement structure. It is clad with a Photovoltaic Curtain Wall Facade. The developer anticipates project completion 12 months after beginning work on site.

Think carefully, what types of Partnering strategies could be used in the above case and why?

Consider the Project Partnering approach and answer the following questions:

1. Which companies could the developer invite to form the initial project team?
2. What should be the subject of the initial partnering workshop and why?
3. How many subsequent partnering workshops would you organize and what would be their purpose?

Now consider the Strategic Partnering approach and referring back to the examples of partnering efficiency in the main text answer the following questions:

1. How could the development company implement the Strategic Partnering approach and what could they achieve by adopting this approach?
2. How could the development company implement the seven partnering actions in this particular case?
3. What purposes could the Partnering Workshops serve in this approach?

Further Reading

The following publications are the source of ideas used in this chapter and provide further information for readers.

Bennett, J. and Jayes, S. (1998) *The Seven Pillars of Partnering: A guide to second generation partnering*. Thomas Telford. This guide describes an early stage in the development of partnering in the United Kingdom. It is a sequel to the influential *Trusting the Team* by the same authors. It codifies research into the first attempts at partnering which were guided by *Trusting the Team* and provides early evidence of the benefits of project and strategic partnering.

Bennett, J. and Peace, S. (2006) *Partnering in the Construction Industry: A Code of Practice for Strategic Collaborative Working*. Butterworth-Heinemann. This guide to best practice was written in cooperation with the Chartered Institute of Building. It provides a detailed guide to project partnering, strategic partnering and strategic collaborative working. Each stage in the development of the most effective forms of partnering is illustrated by case studies of successful practice.

Bresnen, M. (2009) Living the dream? Understanding partnering as emergent practice. *Construction Management and Economics* 27(10), 923–33. This paper describes important research which makes clear that partnering requires more than a predetermined set of actions which encourage cooperative behaviour. Individual projects are influenced by local practice and the particular tools and techniques used by the construction teams involved. As a result projects develop distinctive trajectories shaped by the teams and individuals involved. These situational factors influence the success of partnering every bit as much as the sets of actions described in best practice guides.

Chapter Nine
Total Construction Service

9.1 Introduction

The highest levels of efficiency are achieved by companies providing a total construction service. They enter into contracts with customers to construct new facilities and provide support services. In doing so they aim to mimic the experience of buying a new car where the manufacturer provides a wide range of clearly defined options to allow customers to satisfy their individual needs. New car customers are offered a range of financial options, servicing and maintenance packages, and help with learning how to use their new car most effectively and economically. The car is handed over to the customer exactly at the predetermined time and date, at the predetermined, fixed and very attractive cost. It is sparkling and everything works beautifully; and there is a bottle of Champagne and bunch of roses in the back of the car to help the customer celebrate. Companies providing a total construction service believe buying a new constructed facility should be a similarly enjoyable experience for their customers.

Achieving everything the world's greatest car companies provide for their customers is difficult. It has taken decades of relentless innovation, development and attention to detail to achieve their world class efficiency, profitability and customer service. Regrettably it is unusual for construction companies to even begin to approach this level of achievement. Most construction professionals see construction as fundamentally different from manufacturing and therefore it is obviously impossible for construction companies to behave like the major manufacturing companies. However, there are cases which demonstrate this view is wrong and construction can provide its customers with world class products and services.

9.2 Total construction service providers

The total construction service is provided by several kinds of companies. The largest of these result from organic growth over many decades of building up their experience in a diverse portfolio of projects. They have grown into large corporations employing tens of thousands of people working in numerous divisions and subsidiaries. These companies work across many different sectors and are capable of undertaking the most challenging of construction

Construction Management Strategies: A Theory of Construction Management,
First Edition. Milan Radosavljevic and John Bennett.
© 2012 John Wiley & Sons, Ltd. Published 2012 by John Wiley & Sons, Ltd.

projects, including regional energy and infrastructure networks as well as buildings of all types and sizes for private and public sector customers.

These companies actively develop internal specialist core teams able to plan, deliver and operate constructed facilities. These multi-discipline teams include designers, construction managers and facilities managers. They work with directly employed multi-skilled construction teams that undertake key production and commissioning actions. External companies are appointed only for minor and often very specialist works. Subcontracting is used to employ a relatively large number of specialist component suppliers who work on a long-term basis in partnership with the total construction service provider. These arrangements enable effective teams to move from one project to another with only minor changes to the core team membership. To establish and retain a high level of control over their projects, these companies employ and invest in their workforce so employee turnover is low and their whole workforce is well trained.

Most major total construction service companies have their own research and development departments. These exist in various forms designed to deliver innovative solutions and assist in solving project problems. Some even develop their own construction plant and equipment. The research and development departments are often organized as distinct profit centres that in addition to serving internal project requirements, work for external customers in order to compensate for slack periods between major projects.

In addition to the very large companies, there are many far smaller total construction service providers who operate in niche markets. A number of these companies deliver high-quality prefabricated structures providing, for example, housing, factories or warehouses. They have developed highly automated means of production with components being produced in factories and delivered just-in-time to construction sites where highly skilled teams of directly employed workers assemble the components in a fraction of the time normally needed for a traditionally constructed facility. These relatively small companies employ workforces that are impressively skilled at all levels and as a result their business strategy can rely heavily on their innovative capacity. Companies of this kind continuously improve their internal processes as well as production techniques, services and products. Many are independent family run businesses which comprise a group of interlinked companies that jointly deliver the total construction service. They work in a completely integrated way in which in-house specialist teams move together from one project to another in a synchronised fashion based on deep mutual understanding. As a result these companies are characterised by operational consistency, skilled teams undertaking well understood actions, and high levels of sophisticated automation in every aspect of their work.

Total construction companies of all sizes tend to optimise their efficiency and minimise the number of involved teams through multi-skilling where individual teams can complete a whole series of seemingly unrelated activities that run in sequence. Such a level of optimisation is only possible with highly experienced teams that have gradually extended their skill base through rigorous training regimes being applied over many projects. It is therefore common for total construction companies that rely on automation and multi-skilling to have a skilled workforce which is far superior to the construction industry norm. Their employee retention rates are also far higher than their more traditional competitors can achieve. This all requires considerable investment, which can only be justified by continuous innovation aimed at establishing and maintaining the superiority of their products and services. This inherent strength cushions these companies against economic downturns far better than focusing on the project flexibility exercised by construction companies following the strategies described in previous chapters. Unlike project-focused construction companies, total construction companies nurture their resilience by continuous company-level learning aimed at innovation leading to high-tech automation and total prefabrication. The examples provided in the following paragraphs and the case study in this chapter provide insights into the operational strategies of total construction companies.

9.3 Industrialised housing

Many countries have used industrial methods to provide housing during periods of high demand since the middle of the twentieth century. The most significant examples are in Japan where major conglomerate companies invested in industrialised housing using factory-based prefabrication methods. They adapted the methods developed in Japanese manufacturing industries to produce high-quality, individual houses which they market as products backed by sophisticated customer services.

Japanese manufacturing developed lean production. This is the generic name for linked systems which enabled production to use less of everything than United States mass-production techniques. The driving force is market research to establish and understand customer needs. This guides research and development in product and process engineering and cooperation with suppliers in joint technological developments.

The key outcomes are flexible machines operated by skilled workers which can make a range of different parts as they are needed. This eliminates mass-production's need to store large numbers of parts and improves quality control as the skilled workers are closely involved in controlling their machines and so are immediately aware of defects. Parts are supplied to production lines just-in-time. In effect each step in production calls up parts just as they are needed. A key initiative in ensuring these systems work effectively is the establishment of quality circles. Work teams are trained in problem solving and required to actively search for improvements to their own performance. Each work team has a quality circle meeting every week to discuss their work and find ways of improving their own performance. Twice a year every work team has to report on the most significant improvements they have put into effect in the previous six months. It is common for major companies to hold annual competitions in which the improvements are judged and the best given an award at a formal ceremony which shares some of the characteristics of an Oscar Ceremony.

These various initiatives are linked as quality circles often consult specialist departments within their company, including in particular the research and development institute, as they devise more effective ways of working. This regular interaction provides a company-wide forum which, ultimately, is aimed at finding practical ways of responding to new demands from customers which may be identified by feedback from projects or wider market research.

Lean production is developed further by redesigning products so different models used standard parts and sub-assemblies. These fit together easily in different ways to produce a huge variety of distinct products.

This developed system provided the model for major conglomerate companies as they establish new companies to manufacture industrialised housing. They aim to produce high-quality houses for middle to high income customers. They offer a wide range of design options which give customers considerable choice in the size and layout of their house, its external appearance, internal decoration and fittings, and the comfort and convenience systems it provides.

The companies have sales offices in all of Japan's major urban centres where customers can discuss their needs in comfortable and convenient surroundings. Key issues for the company are the availability of a suitable site for a house and the customer's financial status. As the discussion progresses, serious customers and their families can visit regional show rooms. These display all the options available for every part of their house. Customers can walk up and down the thirty or so available staircases, open and shut doors and windows, try the many different bathrooms, play with kitchen layouts, and experience the character of bedrooms and family rooms. They can discuss technical issues, see the effects of natural and artificial light and ventilation, consider safety, comfort, convenience and environmental issues, and generally become excited at the possibilities offered by modern industrialised construction methods. While these design considerations are underway, the company's staff appraises the customer's site to ensure they can make confident decisions about quality, time and cost.

Once all the decisions about a house are made and the price and completion date agreed, an efficient production system is set in motion. All the parts and sub-assemblies needed to assemble the house on site are called up automatically. This starts a process in which suppliers deliver parts to the company's factories where they are assembled into prefabricated components. Work on site begins by clearing any existing structures and producing the foundations and underground services. Once this essential site preparation is complete, the prefabricated components are delivered just-in-time to be assembled by an experienced production team. Every stage is subject to strict quality control leading to the handing over of a fully complete house exactly on time. It usually takes a little more than three months from the date the sale is agreed to the family moving into their new home.

The major industrialised housing companies invest in training colleges for their own staff. The colleges run courses aimed at helping everyone in the company achieve their potential in ways which increase customer satisfaction and the company's performance. The companies also run courses at their own colleges for their suppliers. These courses provide instruction in site safety, quality control, specialist knowledge and skills aimed at improving site assembly techniques, and good manners in dealing with customers, neighbours and local communities.

The major industrialised housing companies undertake well-funded research into understanding the way their houses are used by different types of adults, children, senior citizens and pets taking account of a range of mental and physical capabilities. This guides the design of every part of their products and the way the sales force deals with potential customers. They study power generation and storage, energy consumption and all the other environmental impacts of the manufacture and use of their houses to devise ways of ensuring they have a minimal impact on the environment. They research the chemical concentrations in their houses to ensure they are minimised and have no damaging effects. They research every aspect of production to provide the basis for steadily improving quality and efficiency. They look for ways to use eco-friendly and sustainable materials wherever possible. Their research and development institutes aim to produce houses which bring joy and foster healthy lifestyles.

The companies hold joint seminars with groups of customers to discuss ideas for improving their products and services. One outcome has been to provide expert advice to customers wanting to re-model their houses as their needs change. The companies also provide financial support for organizations aiming at sustainable communities, those that help disabled people, and schools.

The key challenges facing these highly impressive industrialised housing companies are: ensuring a steady demand for their products; balancing the efficiency benefits of standardisation with customers' demands for variety in their houses; variable and rising transport costs; and the management of their dispersed site assembly teams. There are a number of industrialised housing companies in Japan, including Sekisui House, Daiwa House and Misawa Homes, which have been in business for about half a century. They demonstrate that construction can meet the key challenges and find ways of using modern manufacturing and marketing to provide their customers with a total construction service. Figure 9.1 summarises the overall strategy adopted by these companies.

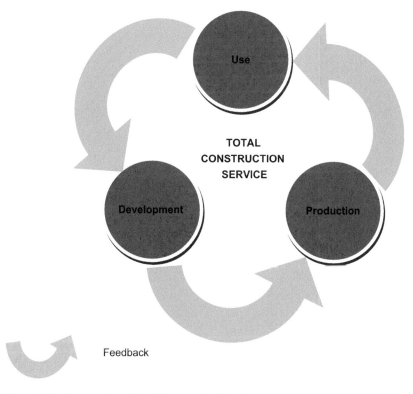

Figure 9.1 Strategy for a Total Construction Service.

9.4 General construction

Housing provides the largest construction market for reasonably consistent products which is why it supports the most highly-developed cases of companies providing a total construction service. General construction faces demand for a significantly greater variety of products which, in most situations, results in a fragmented construction industry. This forces customers to deal with several construction companies or accept a limited level of service. An important exception is provided by Japan's 'big five' construction companies. Kajima, Obayashi, Shimizu, Taisei and Takenaka provide their customers with exceptional levels of reliable performance and service. They provide clear examples of a total construction service aimed at a general construction market.

The 'big five' companies provide a single-point of contact and responsibility between their customers and the construction industry by means of a sophisticated design build approach. Over many decades with government support they have developed into companies providing high-quality products and services. They now achieve all this reliably and efficiently. There are other construction companies in Japan that challenge the leadership role of the 'big five' by undertaking projects all over the world with the same level of unparalleled efficiency. For instance, the Prudential building in Reading, England depicted in Figure 9.2 was built by Kumagai Gumi, a major construction conglomerate from Japan. The building was built to the highest standard and even after several decades of use it shows very little signs of ageing. Unlike many other modern office buildings it radiates quality, durability and is already one of the landmarks in this major Berkshire town.

The approach of Japan's impressive construction companies is outlined in Chapter 1 in the section 'Contractor-led practice.' It is now appropriate to describe their approach in more detail.

Figure 9.2 Prudential Building in Reading, United Kingdom Built by Kumagai Gumi, a Japanese Construction Company.

Secure demand

The companies have invested long-term in steadily improving their knowledge and skills, and developing the systems which enable them to use their strengths profitably. The essential foundation for this is a reasonably secure demand based on long-term relationships within enterprise groups and a guaranteed share of major government construction projects.

An enterprise group or family is a group of companies that cooperate. Each family includes companies from all the major sectors of the economy, including construction. The families compete with each other for market share. However, inside each family, companies work together. It is therefore normal for a major mining, manufacturing, financial, retail or communications company to employ the construction company member of the family to construct all their new facilities. The basis for these projects is usually incredibly straightforward. The customer tells the construction company what they want in the new facility, how much they will pay for it and when it must be complete. This direct approach is possible because Japan has published information about every aspect of the construction industry's work. Customers know what can be achieved and so they can make challenging but achievable demands on their construction family member. It is unthinkable for any major company to deal with any other construction company. This deep cooperation allows companies to share information, support each other and ensure they all work to the highest standards. Competition between families is fierce and provides a vital driving force for continuous improvement and innovation.

The 'big five' provide leadership for Japan's construction sector in dealing with government. They represent the industry's interests in discussing new legislation, economic policy and any other issues being considered by government which may impact construction. They invest in large, well-equipped research institutes, adopt best practice management methods and train their own staff and their established subcontractors in quality control, safety and the use of quality circles to relentlessly search for better ways of working. They publicise successful innovations so they can be adopted by smaller construction companies. In return the government gives all major construction projects to the 'big five.' They bid in competition with each other for these projects but the outcomes consistently represent a 'fair' share for each of these leading companies. Indeed many of the largest projects are undertaken as a joint venture between two or three of the 'big five' companies.

Quality and efficiency

The carefully fostered support from family companies and government gives the 'big five' sufficient confidence to focus on high quality and efficiency. This starts with the company recruiting engineering graduates from Japan's top universities. These highly-educated individuals expect to stay with the company for the whole of their working life and to rely on the company pension when they retire. The 'big five' in recent years have recruited some women but the vast majority of their employees are university educated men. New recruits start with an induction course which takes two to four weeks and teaches the general principles of company practice. Subsequent training is 'on-the-job' beginning with six months of watching the activities of various departments and visiting construction sites. They are then allocated to a department where they learn the company's knowledge and skills. Every six months their progress is evaluated by their immediate superior in a formal discussion of their aims and ambitions. Promotions depend on these evaluations. Anyone judged to be performing badly is moved to another department where it is hoped they will do better. A few employees are moved several times before they find a department they can work in effectively.

Workers at all levels are expected to search for ways of improving established methods by meeting every week in quality circles. Innovations are developed in the company's research institutes. This controlled approach to steady improvement and innovation enables the companies to combine reliable efficiency with change. When a better way of working is identified by ideas developed in a quality circle or the research institute, it is discussed widely by those likely to be affected. Only when there is wide understanding of the change is it introduced. Then it is adopted as part of the company's standard approach until a new and even better idea is found, tested and agreed. This approach ensures steady, controlled improvement which over decades delivers world class performance.

As young workers learn all of this, their on-the-job training is supplemented by occasional courses covering specialised technical matters or management issues. Employees are moved to new departments every two or three years to give them a wide knowledge of the company's work. They are given the formal rank of Deputy Manager of a section by the time they reach the age of thirty-five. Subsequent promotions are to Manager, Deputy General Manager, and at about forty-five to General Manager of a department. Beyond this are Directors, Managing Directors, Senior Managing Directors, Vice-Presidents and at the top, the President. Promotion largely depends on age but good performance is rewarded by somewhat faster promotion. In this way effective workers reach senior positions late in life having a deep understanding of their company.

Brief

Individual projects begin with the customer's statement of their requirements. This is likely to result from discussions between the customer's organization and the construction company's sales staff. Indeed the 'big five' have members of their sales force permanently based in their major customers' organizations actively searching for opportunities where construction could contribute to a customer's business. When a customer begins to consider the possibility of investing in a new facility, their family construction company makes all their resources available to ensure every possibility is explored and tested before final decisions are made. As a result, the customer's statement of their requirements is extremely detailed and often includes drawings and specifications of exactly what is to be produced. Occasional customers from outside the family group may employ consultants to help them decide the details of their new facility. Others are content to provide simple, generic descriptions. Irrespective of the status of the customer or the amount of information provided, the construction company appoints one of their senior, experienced managers to take responsibility for the project. They ensure it moves smoothly through the construction company's departments. The required facility is handed over to the customer fully complete, exactly on time.

Design and plan

This totally reliable performance results from extremely detailed design and planning of every aspect of the project. This is produced under the direction of the project manager by experienced engineers with the help of highly developed systems which embody the company's standard design details and construction methods. Prefabricated components are used wherever possible to reduce site activity to the minimum consistent with meeting the customer's requirements efficiently. Drawings show every element and every part of every element. Specifications are largely based on comprehensive national standards which mean project specifications can be extremely brief. Every action required to complete the project is planned in detail. For example, every day's production on site is planned. This means all the subcontractors subsequently involved in the project know exactly what their construction teams will be doing day by day. It means those responsible for providing and operating plant and equipment on site know exactly what is required. It also means companies supplying materials and components know the precise date and time they are to make their deliveries to the site.

The subcontractors needed to undertake the manufacturing, production and commissioning are selected by the project manager. All the companies have worked together for decades, in some cases for centuries. This allows the project manager to tell the selected subcontractors what they must do, when they must do it and how much they will be paid, confident this will be all be accepted. The subcontractors trust the construction company to ensure they have a regular supply of profitable work. These arrangements are simple and efficient and rely on the technology and production techniques used being familiar to everyone involved. Any departures from established methods will have been thoroughly tried and tested, discussed with the subcontractor, and any necessary training provided before innovations are used on a project.

A common criticism of this disciplined approach which is often heard in Western countries is it is overly mechanistic reducing workers to mere machines. However the opposite is nearer the truth, the high level of stability and certainty enables employees to relentlessly search for improved ways of working which results in a continuous stream of incremental improvements. It is the West's overly competitive approaches which force workers to concentrate on day-to-day fire-fighting which is seriously inefficient. The continuous search for ways of improving information, tools and materials is found only in total construction companies and is almost entirely absent everywhere else throughout the construction industry.

Manufacture, production and commissioning

Once production begins, the project manager, supported by an experienced team of engineers from the construction company, is based on site full time. He is totally responsible for ensuring manufacture, production and commissioning are carried out in accordance with his design and plan. This is achieved by means of a consistent pattern of work on site every day. It begins at 8.00 a.m., but workers have been arriving on site since soon after 7.00 a.m. They change into their working clothes and are lined up in construction teams ready for work at 8.00 a.m. The day begins with 10 minutes of exercises to music played over the national radio. The exercises are learnt at school and practiced daily throughout Japan. Then the project manager flanked by his team briefs the workforce on the day's work. The briefing describes the main production activities, deliveries and any safety issues needing particular attention. Then each construction team moves to its work space and holds what is called a 'toolbox' meeting. This is run by the team's foreman and describes exactly what the team must achieve before the end of the day's work and exactly how they will achieve that target.

Work continues, with a break for lunch at 12.00 noon, until 5.00 p.m. However if the team has not completed its day's work by that time it continues working until it has met its target, no matter how late that turns out to be. The key control over this steady, orderly pattern of work is the coordination meeting which takes place at 3.00 p.m. every day. All the subcontractors' foremen and the project manager with his team meet to review progress and remind themselves of the next day's work. Any problems are discussed and everyone concentrates on finding a solution. Anyone can make suggestions which are discussed in turn until an effective way forward is identified. This is announced by the project manager and everyone accepts his decision. Then the next day's work is discussed. This is defined by the design and plan produced before production on site began so the meeting is merely checking that every foreman knows exactly what their team must achieve. The coordination meetings are impressive in the way everyone concentrates on ensuring work is undertaken exactly as originally planned. There is no discussion of contractual issues or claims, simply a relentless focus on completing the project on time. The meetings usually take no more than 20 minutes to solve any problems and reach total agreement on the next day's work.

Even with this deep cooperation by very experienced teams, difficulties arise. To allow for this the plan includes some slack. Each production activity is planned to be undertaken by one construction team working one shift throughout each normal week. This means a project can be brought back on schedule by teams working on into the evening or over a weekend, bringing in a second construction team to work a second shift, or in extreme cases bringing in more than one extra team working in parallel. Whatever it takes, the project is returned exactly to the original plan at the earliest possible date. In no circumstances is a project ever finished late. That would be a matter of extreme dishonour and a commercial disaster for any major construction company.

Quality and safety

Throughout all the production activities, quality control is given the greatest attention. Japanese customers expect new things to be perfect and work properly. Quality begins with the choice of tried and tested design details, components and systems. On site, quality is measured continuously using predetermined tests which specify a detailed test schedule for every part and element of the new facility. The results of the tests are recorded in photographs which are stored next to photographs of the work which is the subject of each test. The procedures which guide this total quality control system are set down in manuals produced by the construction firms. They ensure new facilities work exactly as intended when they are handed over to the customer.

Safety is taken just as seriously as quality. It is designed and planned into production methods. Provision for safe access is prefabricated into structural components. Sites are rendered inherently safe by enclosing the whole perimeter of tall structures with sheet steel and plastic protective screens fixed to the scaffolding. This provides protection from the weather and greatly reduces the risk of workers or objects falling off the structure. Materials are stored in clearly marked areas and their distribution is planned with safety in mind. Sites are rigorously kept clear of rubbish and time is allocated every day to ensuring the site is tidy. The aim is to ensure workers do not face unexpected and therefore potentially dangerous situations. Safety slogans and special posters are displayed at key points all around construction sites. All this is checked as part of regular inspections by the construction company's safety inspectors.

Quality and safety are relentlessly emphasised in all the training provided for workers at all levels. This focus is accepted as an essential element of efficient construction. The need to re-do work, correct defects and deal with accidents undermines even the most careful plans. Therefore these major causes of variability are eliminated. Everything possible is done to ensure production on site is smooth and orderly.

Quality circles

Quality circles allow workers to discuss ideas for improving their work within a formal structure. The most promising ideas are considered in detail at subsequent quality circle meetings. A quality circle can ask the company's research institute to undertake research which helps develop an idea. When a new way of working appears to provide benefits, it is discussed with other workers likely to be affected. If a consensus in favour of a new idea is established, the quality circle plans its introduction and puts their innovation into practice. In this way quality circles ensure a steady stream of practical improvements to the company's performance.

Research institute

Big ideas, design concepts, new technologies, improved engineering or a more efficient construction method are considered by engineers and scientists at the research institute. Most of their work is concerned with tackling advanced engineering problems. Considerable effort is devoted to ensuring the new facilities they construct will withstand Japan's frequent earthquakes. Other high-priority work is aimed at reducing the environmental impacts and energy use of construction actions and the facilities they produce. Similarly research into ensuring new facilities are safe, comfortable and generally people friendly is regarded as important. Research into construction techniques consistently seeks to improve quality and safety. It also seeks to reduce the amount of waste produced, look for ways of recycling resources and ensure facilities are constructed in ways which are easy to alter or demolish. Beyond this directly practical research, the research institutes make proposals for incredible and exciting developments. Cities under the sea, buildings one-kilometre high capable of housing whole communities, automated production on site using robots, materials based on biotechnology, and many more ideas are developed and publicised in the national media. This not only helps build the reputation of the individual company but also helps maintain construction's public image as a modern, technologically-based industry.

9.5 Other total construction service companies

The market conditions engineered by Japan's leading construction companies are difficult to achieve. Nevertheless interesting moves towards developing construction companies able to provide a total construction service are being made in other countries. For example, companies aiming to provide a total capability are starting to win an increasing number of projects in Japan's great Asian rival, China. The approach is often called design build but the deeper integration of construction actions which characterises the total construction service has been identified in research. The research has identified the competences needed by successful companies as: the ability to coordinate the activities of all the different types of construction teams; establishing and maintaining competitive strength in the construction market; the ability to develop new products and services especially those based on technological competence; financial security; highly qualified personnel; and a good reputation among customers. These characteristics are leading the big Chinese companies towards the same all-round competence characterising their Japanese rivals.

In looking for examples of companies providing a total construction service in Western economies, it is sensible to look at central Europe but in particular Germany which consistently comes close to matching Japan's performance in international comparisons of manufacturing and construction performance. The following case study describes an excellent example of a German company providing a total construction service.

Case Study: HUF Haus GmbH

HUF Haus GmbH (HUF) is a family run business involved in commercial and residential projects with headquarters in the village of Hartenfels, northwest of Frankfurt. The company was founded in 1912 by Johann Huf initially as a carpentry business centred on a sawmill. The company widened its business interests, grew steadily and gradually gained nationwide recognition as a high-quality manufacturer. For example, they built the German and Arabic pavilions at the Brussels World Fair in 1957/58. This helped give impetus to the development of their first prefabricated timber structures which culminated in the HUF Kaufhof house. The company perfected their house designs through a series of progressive steps begun in 1964 with the advent of HUF 'Ideal House' and followed by 'Timber-Framed House 2000' in 1972, 'House Generation' with a double-pitched roof in 1992 and finally the 'Green[r]evolution House Generation' in 2009.

Into the 1970's HUF Haus GmbH subcontracted a large proportion of their work but they gradually realised subcontracting provides a far lower level of control over the work in progress. This led the company to slowly consolidate their operation through a series of carefully crafted acquisitions

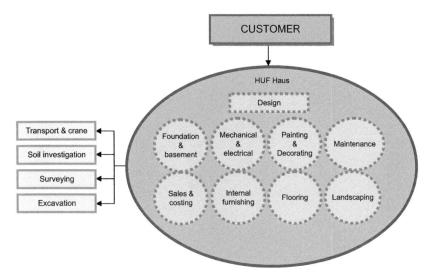

Figure 9.3 HUF Haus GmbH Project Organisation.

of their subcontractors. The acquisitions were smooth and did not affect the operation in any way because their working relationships with the specialist companies had been built over many years. It was efficient to consolidate their businesses because by itself this brought a number of benefits culminating in far greater control of the total construction process. As a result HUF Group now comprises a number of specialist companies, each profitably operating in its own market segment. Each specialist company has almost total control of the construction process and as Figure 9.3 shows only excavation works, transport, the provision of cranes, soil investigation and surveying are outsourced because they represent a relatively minor part of the process. However, even these activities are subcontracted to companies with whom HUF maintain very close relationships spanning over many years.

The company is constantly growing, both in the level of services provided as well as geographically. Their investment decisions are predominantly driven by the search for improvements and the continuous development of products. The most innovative aspect of their business is the product itself and their ability to provide customers with a price and specification at an early stage of a project no matter how complex the requirements. In the more sophisticated projects, the company's specialists meet the customer to discuss, test, agree and define the individual requirements.

The construction of a new house begins with a customer and HUF's architect agreeing the design of the house. The customer pays a 10 per cent fee upfront for architectural services. This is refunded once the fit-out protocol is finalised. The rest is paid in a series of instalments corresponding to various stages of the project: 30 per cent after the fit-out protocol is finalised and signed; 20 per cent after manufacturing; 25 per cent at the start of assembly on site; a further 20 per cent just before the full fit-out begins and the final 5 per cent at completion. The architect is directly employed by the customer with the aim of ensuring an individually tailored service. This may require a number of discussions which continue until the

Figure 9.4 Ready-Made Building Components Awaiting Final Painting in Order to Avoid Damage During Installation.

customer is able to agree all the design requirements, although this is a relatively straightforward process since HUF work on a set of standard designs and components. Once the design has been agreed the HUF team prepare contract documents, hand them over to the customer along with the door key. The high level of standardisation allows HUF to immediately proceed with the planning application process, which includes all required surveys (e.g. soil, site, traffic, etc.). When the planning application is approved HUF invite customers to fully specify the fit-out at their premises in Hartenfels in Germany. The fit-out procedure normally takes about three days and once completed HUF teams produce and issue a so called fit-out protocol, a document that includes all the details and costing for each individual item in the finished house. The customer then needs to approve the fit-out protocol in order to start the manufacturing process. Performance of the new facility is set out within the contract as well as within the building specifications and fit-out protocol so there are no surprises. The customer is allowed to change the design up until the start of manufacture. From this point it takes about 4 months for the house to be delivered to the site, and the assembly of the structure takes about a week. Figure 9.4 shows ready-made components are delivered on site precisely when required and installed using often specialised custom-made tools and equipment. The components are made of materials that have been developed in-house and are innovative products in their own right but they are available only to their customers in order to maintain a competitive edge. The structural assembly is followed by the first fit, finishing and furnishing, and approximately one year from the initial discussion with the architect the house is handed over to the customer. HUF Haus GmbH are involved in high-quality private residential projects as well as commercial and healthcare projects but the above process is applied in all their projects regardless of the size

Figure 9.5 German Commercial Building Built by HUF Haus GmbH.

and function. Figure 9.5 shows an example of a commercial office building in Germany.

In relation to project management, HUF dedicates one contact person for each stage of the project to maintain their relationships with the customer. However, time, cost and quality are controlled by the fit-out manager during the preproduction stages and the site manager during production. A post-construction review is carried out by an independent control department. The frequency of project meetings depends on the project size but usually there are dedicated weekly meetings on site ensuring that every stage is delivered according to accepted specifications. In addition the site manager visits the site a minimum of seven times per week or as and when required. This overlap of meetings is of particular importance because site activities are interlinked with manufacturing. While site conditions may dictate the time of production, manufacturing facilities with limited capacity dictate the production programme. Therefore both are managed by a single department.

The company predominantly works with new customers so it is important the above procedure is standardised in order to avoid confusion and delays during construction. HUF attract new customers with the help of display houses, which are used to demonstrate the quality of the building, including the design quality, material quality and end finish quality. They offer their customers sophisticated after care services which may include a full lifetime maintenance package.

9.6 Total construction service efficiency

The efficiency of the total construction service approach derives from: embodying best practice in company standards; the relentless application of those standards on individual projects; and the steady, systematic search for ever better ways of working. The resulting strategies include many examples of actions which support all the construction management propositions as the following descriptions illustrate.

Simplify

Total construction service companies take many initiatives which reduce the number of individual construction teams involved in construction projects. These begin with the creation of a core team for each project inside the company. Site activities are simplified by a steadily increasing use of prefabrication and even automation on building projects. Attempts by Japan's 'big five' to eliminate construction teams from production by using robots on completely automated construction sites failed on economic grounds. Trials of the technologies demonstrated that automation can produce complex buildings but the cost was prohibitive. However, the research into robotics produced many spin-offs which are now used in construction plant and equipment. In particular, major civil engineering projects increasingly use highly sophisticated machines which approach the capabilities normally associated with automation to reduce the number of individual construction teams involved in production actions.

Almost every aspect of the approach adopted by total construction service companies serves to minimise the intensity of the interactions between the construction teams involved in individual projects. The core team for each project produces an incredibly detailed design and plan to provide simple, clear directions for the other teams involved. Problems are dealt with immediately at project meetings. These occur regularly throughout the all stages of the project and focus absolutely on solving problems. Problems are not allowed to fester and as a result interactions between the production teams involved are minimal.

Certainty

The careful preparation of the customer's statement of requirements provides a robust basis for core teams to run their project efficiently. They, with the selected subcontractors, consistently use the company's standardised answers in preparing the extraordinarily detailed design and plan. Manufacturing, production and commissioning activities are carried out exactly in accordance with the design and plan. This serves to reduce the variability delivered by construction teams. To help ensure this, production plans provide slack time which can be used to compensate for any delays which do occur. This is sufficient to ensure that every project sticks to its production and commissioning plan. The consistent result is that every construction team on site carries out its predetermined work every day. In other words, variability is reduced to virtually nil.

Total construction service companies go to considerable lengths to protect construction teams from interference. This begins with the careful evaluation of the customer's requirements before the project team is set up. As a result, it is unusual for a customer to ask for changes once the project is underway. In Japan the close links between major construction companies and government have many benefits not least in insulating individual projects from bureaucratic interference. As far as possible, production actions are protected from the effects of bad weather. In Japan this includes the provision of robust protective screens around building structures as they are produced on site. Total construction service companies tend to work through long-term relationships and foster a culture of meeting all their obligations. This helps ensure activities can be planned in detail with every confidence of exactly putting the plan into action; and ensuring the customer's completion date is achieved.

Construction teams

Japanese culture ties individuals to their companies throughout their working lives. This maybe extreme but most total construction service companies focus on retaining their staff. This makes it sensible to take every opportunity to improve the quality of support provided to construction teams. Such support represents an investment in the company's future. Also total construction service companies actively ensure their regular subcontractors provide excellent support for the construction teams working on their projects. This includes training, especially in quality control and safety, thorough briefing of teams before they arrive on site and in identifying ideas for improving their products or services. These important actions are encapsulated in Japan's use of quality circles.

The policy of investing in the competence of construction teams is reinforced by their active involvement in the steady development of standardised design details and production methods. This, together with extremely detailed design and planning based on the use of these standards ensure all construction teams are competent in the technologies they are required to work with. All this is solidly based on the long-term relationships between customers, construction companies and their subcontractors.

Everyone involved in providing an effective total construction service works on the basis of agreements recognised as fair. They provide the glue which sustains long-term relationships. An important benefit is construction teams fully accept the agreed objectives of the project of which they form part because they know from experience they are in their own best interests. More than this they are motivated to achieve the agreed objectives because they know the success of all their interlinked companies is related to the efforts of everyone involved.

Interactions

Well-developed total construction service companies work through tightly integrated systems directed by the core team. One benefit of these efficient systems is there is little need for communication between the construction teams involved in a project organization and that which does take place is in formal meetings chaired by the project manager or a member of his core team. This also helps ensure accurate communication between the construction teams involved in a project organization.

It is normal for Japanese customers to tell construction companies what to produce and specify the terms and conditions which will apply. Similarly construction companies tell their subcontractors what they are required to do on what terms. It is difficult to envisage any way of further reducing the negotiations needed to agree the transactions which bring construction teams into a project organization. A key to ensuring these arrangements are accepted is the publication of extensive data about every aspect of Japanese construction. This means everyone involved in agreeing contracts is well informed. People know they have a fair deal and inevitably this serves to ensure construction teams regard the transactions which brought them into a project organization as advantageous to themselves. As a result there is no temptation to waste resources attempting to improve the terms of the transactions.

This virtually automatic approach to transactions is unique but all effective total construction service companies devise straightforward ways of forming fair agreements with their customers and then ensuring they deliver exactly what the customer expects. An important reason for this care is they understand the importance of having a good reputation for treating customers well.

Market response

The essential basis for the long-term investment which distinguishes the approach of the most effective total construction service companies is a secure market. All these successful companies invest huge amounts of time and effort in ensuring a steady demand for their products and services. They totally recognise it is fundamental to their survival that they satisfy the requirements of local or specialised construction markets in ways which ensure their company's long term survival.

Development

Their long-term approach allows total construction service companies to invest in ensuring construction teams have well developed skills and knowledge which match the requirements of the construction projects they undertake. This applies to their own staff and the subcontractors who undertake all the manufacturing, production and commissioning activities on their projects. The bases for this deep competence include: technical education followed by on-the-job training; specialised training courses; and in Japan, universal membership of quality circles. This all serves to ensure construction teams are integrated by well-developed internal relationships and project organizations behave exactly as if everyone involved belonged to the same one organization.

Innovation

Total construction service companies' investment in innovation is unmatched. For instance, in Japan the 'big five' have large, well-equipped and staffed research institutes. These work closely with quality circles which actively search for potential improvements to products and services. This link ensures the innovations produced by the research institutes match the requirements of the construction projects their company undertake. This close connection between every construction team and world-class research is not achieved in any other national construction industry.

All total construction service companies actively search for improvements to the way they work. A crucial part of the systems which identify, develop and test new ideas and innovations is that this is not allowed to compromise individual projects. Ideas for better ways of working or innovations which result in improvements to facilities frequently emerge from projects but they are not used on those projects. They are recorded so they can be properly considered and evaluated. Promising ideas and innovations are developed and tested to ensure they provide real benefits and are consistent with the company's standards. When it is fully developed, the new idea or innovation is tried on several projects. The construction teams involved are given any necessary training and the application of the new idea or innovation is monitored by the research institute. This lengthy and careful process delivers steadily improving performance, more highly developed facilities and no surprises for manufacturing, production or commissioning.

Companies

Total construction service companies ensure their organization has effective information systems by standardising the information which needs to be communicated and using consistent patterns of meetings on all projects. More general communications inside companies are effective because everyone learns how the information systems work as an integral part of their training. In Japan, this ingrained understanding is continually reinforced by a culture in which significant decisions are based on company-wide discussions. Indeed most major decisions have their origins in middle-management debates and wide consultation. Ideas flow up through Japanese companies. This is in stark contrast to the Western idea of senior managers providing leadership and being decisive.

It is often observed that the real difference between the two approaches is that in Japan, decisions are debated and discussed before the decision is made while in the West the debate begins only after a senior manager has announced a decision. There is some truth in these caricatures but one clear and significant outcome of the Japanese approach to decision making is construction companies can rely on their organizations' communications being effective and producing common understandings.

A central part of enabling a total construction service company to survive long term is ensuring common values run consistently through every aspect of their organization's work. An absolute focus on meeting customers' requirements, concentrating on quality and time control, ensuring construction sites take safety seriously, and relentlessly searching for innovations and better ways of working shapes everything the most successful companies undertake. This is expressed in an inherent discipline which fosters the habit of discussing issues and problems before decisions are made. This can be criticised for inhibiting individual initiative but it does ensure all the parts of construction companies' organizations act in ways which are intended and authorised.

Company relationships

All successful total construction service companies work through networks of long-term relationships with all the companies that make key contributions to their projects. This is most clearly evident in Japan where the culture of regarding major companies as part of a family is supported by each company holding blocks of shares in the other companies and having directors appointed to the other boards. In this way an intricate web of association is formed which ensures transactions are agreed with the minimum of effort, accepted as fair by all the parties involved, and are acted on in the spirit in which they were agreed. It also ensures the organizations involved in construction projects form relationships with other organizations on the basis of composing since at a deep level they regard themselves as part of the same organization.

Feedback

An integral part of the most highly developed approach to construction projects is the collection of detailed data describing every aspect of production. This is systematically analysed and summarised before being considered by meetings at all levels throughout construction companies. These routine feedback systems support all parts of total construction service organizations. As a result every team in construction companies has feedback about how well they are meeting their objectives and accepts a responsibility for ensuring feedback is acted on. This project control is crucial to each company's success but more importantly it ensures the standards, which determine what is constructed and the methods used, are guided by feedback.

9.7 Construction management propositions

The actions described in the previous section aimed at improving the efficiency of construction companies providing a total construction service are encapsulated in the following propositions provided by the theory of construction management.

To enable construction to be undertaken efficiently within an approach which provides a total construction service, construction management aims to:

- Reduce the number of individual construction teams involved in a construction project.
- Improve the quality of relationships between the construction teams involved in a construction project.
- Reduce the performance variability of the construction teams involved in a construction project.
- Reduce the external interference experienced by the construction teams involved in a construction project.
- Select construction teams competent in the technologies required by the project in which they are involved.
- Ensure construction teams accept the agreed objectives of the project in which they are involved.
- Ensure construction teams are motivated to achieve the agreed objectives of the project in which they are involved.
- Foster accurate communication between the construction teams involved in a project organization.
- Minimise the effort required to achieve accurate communication between the construction teams involved in a project organization.
- Minimise the length and intensity of negotiations needed to agree the transactions which bring construction teams into a project organization.
- Ensure construction teams regard the transactions which brought them into a project organization as advantageous to themselves.

- Minimise the resources construction teams devote to improving the terms of the transactions which brought them into a project organization.
- Satisfy the requirements of local or specialised construction markets in ways which ensure their company's long-term survival.
- Develop construction teams with well-developed skills and knowledge which match the requirements of construction projects.
- Develop construction teams integrated by established relationships.
- Improve the quality of support provided to construction teams by the construction companies of which they form part.
- Foster innovations which match the requirements of construction projects.
- Ensure their organization uses effective information systems.
- Establish values which run consistently through every aspect of their organization's work.
- Ensure all the parts of their organization act in ways which are intended and authorised.
- Ensure their organization's communications are effective and result in common understandings.
- Ensure their organization's transactions are agreed with the minimum of effort, accepted as fair by all the parties involved, foster established relationships, and are acted on in the spirit in which they were agreed.
- Ensure their organization forms established relationships with other organizations.
- Ensure their organization collects, reviews and acts on feedback about the results of the organization's actions on its objectives.
- Ensure their organization collects, reviews and acts on feedback about the effects of established norms and procedures.

9.8 Total construction service performance

The theory of construction management states that efficiency is the extent to which construction organizations achieve agreed objectives. Every construction project undertaken by well-established total construction service companies has the primary objective of meeting the customer's requirements exactly. The fact that this objective is met consistently means providing a total construction service enables these companies to be remarkably efficient. This is especially notable for Japan's major industrial housing companies and the 'big five'.

It is also the case that international comparisons of cost efficiency consistently identify Japan as outperforming other leading national construction industries. This is illustrated in Table 9.1.

The results in Table 9.1 were produced by calculating the cost of six building projects and six infrastructure projects in each country and adjusting the results to purchasing power parity to remove the distortions introduced by market exchange rates. The costing was carried out by experienced practitioners in each country based on detailed descriptions of the projects. The projects were identical across all the countries except where specific details were adjusted to fairly represent normal construction in individual countries. In other words every effort was made to ensure the results measured the performance of the various national construction industries.

Country	Buildings	Infrastructure
Japan	77	88
France	101	67
Germany	96	74
Spain	94	84
Italy	97	92
United Kingdom	100	100
United States	108	112

Table 9.1 Construction Costs

Source: W S Atkins International et al. (1994) *Strategies for the European Construction Sector.* European Commission

Japan's clear superiority in respect of building projects is significant. First building comprises the majority of construction. Second, building involves more distinct technologies giving rise to more interactions which benefit from Japan's cooperative approach. Third, buildings are less influenced by national geological characteristics which are particularly challenging in Japan compared say to France.

In any case the results influenced subsequent thinking across construction management practice and research. The research has been repeated at various dates by research centres in different countries. The results broadly confirm the two biggest surprises at the time the original research was published, namely, the poor performance of the United States and the realisation that Japan has the most cost efficient construction. Other studies suggest the United States can match Japan's performance but this conclusion results from comparing the United States's basic, standardised buildings with Japan's higher quality and ignoring differences in general price levels. Comparing like with like consistently produces the comparative performance shown in Table 9.1.

9.9 The theory of construction management

This chapter provides a thorough test of the theory of construction management. Companies providing a total construction service provide opportunities for all of the propositions to be acted on. The available evidence suggests these companies achieve greater efficiency than any alternative approach. Therefore, it can be claimed that the theory of construction management passes this test and the construction management propositions in the section 'Construction management propositions' are supported.

Example – Building Project using a Total Construction Service

The example relates to new hotel project similar to that used as the basis of the example in Chapter 5. The project is undertaken by a total construction service company which achieves major economies throughout the project compared to all the alternatives described in Chapters 5, 6, 7 and 8. The resulting highly efficient approach is reflected in the following data and calculations. The overall effect is the total construction service project is completed much faster and costs significantly less than the traditional construction approach.

		Number of Teams	Number of team-days
Brief		1	80
Design	Concept design	1	100
	Scheme design	1	80
	Detail design	2	100
	Technical design	4	100
	Production information	2	80
Plan		1	50
Procurement	Select companies	1	40
	Plan and control	2	50
	Contracts and payment systems	1	30
Manufacturing		8	150
Production	Substructure	3	80
	Structure	3	80
	External envelope	3	100
	Service cores	3	100
	Risers and main plant	3	60
	Entrance and vertical circulation	6	200
	Internal divisions	2	70
Decoration		4	250
Fittings		2	200
Landscaping and external services		3	200
Commissioning		4	100
TOTALS		60	2,300

From equation 4.11 there are $R = 1,770$ possible relationships between teams working on this project. The detailed project plan, which can be seen on the website linked to this book, www.wiley.com/go/construction-managementstrategies, shows 1,196 of the possible relationships do not occur. This gives a total of $R-k = 574$ relationships that do in fact occur.

The largest number of teams involved in the same time interval is 14 so there are 91 different relationships.

The largest number of teams that could be involved in the same time interval is 20 in interval 13 so there could be 190 different relationships within that time interval. However, since not all teams work at the same time it is imperative to determine the number of teams that work simultaneously for each time interval. To relate it back to this example, the plan shows during the 13th time interval the following teams work simultaneously:

■ 2 out of 4 technical design teams
■ 2 out of 2 production information teams
■ 2 out of 2 plan and control teams

■ 1 out of 1 contracts and payment system teams
■ 4 out of 8 manufacturing teams
■ 3 out of 3 substructure teams

This gives a total of 14 teams with 91 relationships.

Every team on a project has established relationships with at least one but often several other teams giving a total of 568 established relationships. This gives an Established Relationships Indicator: $E_R = 568/574 = 0.99$ (equation 4.13). This is a very high value which is then further confirmed by equation 4.14 which generates time dependent indicators of established relationships ranging from 0.96 to 1.00.

The relationship fluctuation indicator F_E is calculated using equation 4.15 and gives a value of 0.99. This reflects a very strong influence of established relationships throughout the project because both E_R and F_E are comparatively really high.

However, to really establish the benefits of established relationships we have to look at how long the teams involved are working together on this project and how much time they have spent working together on past projects. Then equation 4.16 gives a Relationship Quality Indicator for each established relationship which ranges from 0.95 to 0.99 and the total Relationship Quality Indicator (equation 4.17) is: $Q_R = 560/574 = 0.98$. The high number of established relationships together with the number of high values for individual Relationship Quality Indicators means the total Relationship Quality Indicator is good because the 568 established relationships out of possible 574 relationships are of very high quality.

According to equation 4.18 we can calculate the relationships configuration complexity indicator C_R:

$$C_R = \frac{1}{3}\left(\left(1 - \frac{30}{119}\right) + \left(1 - \frac{46.7}{1400}\right) + \left(1 - \frac{14}{60}\right)\right) = 0.83$$

The indicator shows the project organization is not overly complex. There are 30 time intervals but there could be a maximum of 119 time intervals. The maximum number of teams working at any given time is 14 in the 13th time interval. The intervals are relatively short in comparison to total project duration, which is beneficial in order to prevent escalation of problems when teams interact for a prolonged period. Nevertheless, relatively higher number of intervals on one hand and a large number of teams working at the same time in one of the intervals on the other counterbalances the above benefits. As a result, when compared to examples in Chapters 5, 6, 7 and 8, the configuration itself is less optimal.

The past performance of the 60 teams over their immediately previous 10 projects provides individual team Performance Variability Indicators which range from 0.95 to 1.0. These give a total Performance Variability Indicator (Equation 4:20): $R_p = 0.99$. This is of course the average performance variability showing the teams managed to complete their work within the agreed time on 99 per cent of their recent projects (note we have assumed c_j to be 1 for all teams although for real life projects these values should be determined separately).

The External Interference Indicator based on historical records for the particular region for the same type of building project undertaken by experienced project organizations using total construction service is 0.84.

Overall, we have the following inherent difficulty indicators which are compared to the indicators for the similar project using traditional construction:

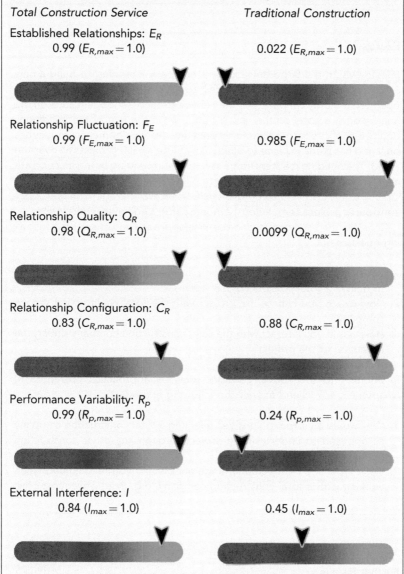

Total Construction Service   *Traditional Construction*

Established Relationships: E_R
    0.99 ($E_{R,max}=1.0$)   0.022 ($E_{R,max}=1.0$)

Relationship Fluctuation: F_E
    0.99 ($F_{E,max}=1.0$)   0.985 ($F_{E,max}=1.0$)

Relationship Quality: Q_R
    0.98 ($Q_{R,max}=1.0$)   0.0099 ($Q_{R,max}=1.0$)

Relationship Configuration: C_R
    0.83 ($C_{R,max}=1.0$)   0.88 ($C_{R,max}=1.0$)

Performance Variability: R_p
    0.99 ($R_{p,max}=1.0$)   0.24 ($R_{p,max}=1.0$)

External Interference: I
    0.84 ($I_{max}=1.0$)   0.45 ($I_{max}=1.0$)

The total construction service project's indicators are very high for Established Relationships, Relationship Quality, Performance Variability and External Interference. These indicators suggest the project has a low level of inherent difficulty. This is reinforced by the high values of the indicators for Relationship Fluctuation and Configuration which show good quality relationships benefit most of the project. There is therefore a high probability the project will be completed exactly as planned.

This reflects the teams' excellent performance resulting from a large number of established relationships which benefit all but six interactions. There is a likelihood of external interference but the project organization's decision to use total construction service approach gives them a realistic chance of completing the project as planned.

Exercise

Imagine you run a large development company intending to build a new 30-storey multi-purpose tower, which is to house an underground externally operated 3-storey car park, and then above ground a shopping zone on the first 5 floors, a hotel on the next 5 floors, while the rest of the building is dedicated office space. Take into account that the development company has built and has plans in place to build many more similar projects all over the world. The building is a steel frame structure surrounding an in-situ concrete core sitting on prefabricated concrete piles and in-situ concrete basement structure. It is clad with a Photovoltaic Curtain Wall Facade. The developer anticipates project completion 12 months after beginning work on site.

Consider the total construction service approach and answer the following questions:

1. What internal structure would a company need to have in place to be able to deliver a total construction service?
2. How would you initiate the project with the total construction service company?
3. How would you interact with the total construction company during the execution of the project?

In addition to the above, consider previous approaches presented in Chapters 5, 6, 7, 8 and answer the following two questions:

1. Why would a customer consider selecting a total construction company as opposed to undertaking a project through any other construction management strategy?
2. What type of a contractual arrangement would you recommend when working with a total construction company (e.g. fixed price, cost plus, etc.)?

Further Reading

The following publications are the source of ideas used in this chapter and provide further information for readers.

Bennett, J. (1991) *International Construction Project Management: General Theory and Practice*. Butterworth-Heinemann. This book describes early research into the approaches used in various national construction industries, including the highly integrated approach of Japan's 'big five' construction companies. The descriptions are used to establish and illustrate a basic theory of construction management.

Bennett, J. (1993) Japan's building industry: The new model. *Construction Management and Economics*, 11(1), 3–17. This paper describes research into the work of the Japanese construction companies' research institutes. It describes these world class research facilities and explains how they support all aspects of the companies' activities.

Gann, D. M. (1996) Construction as a manufacturing process? Similarities and differences between industrialised housing and car manufacture. *Construction Management and Economics*, 14 (5), 1141–52. This paper describes research into the work of the Japanese car manufacturers and manufactured housing companies. It provides detailed information about the factors needed for success in both industries and concludes they both could learn from each other.

Radosavljevic, M. and Horner, R. M. W. (2007) Process planning methodology: dynamic short-term planning for off-site construction in Slovenia. *Construction Management and Economics*, 25(2), 143–56. This paper shows how new planning methodology has been used to improve the performance of a medium-size total construction service company in Slovenia. The research is case-study based and provides a rare insight into a set of specific improvements in a small project being delivered by two different approaches.

Xia, B., Chan, A.P.C. and Yeung, J.F.Y. (2009) Identification of key competences of design-builders in the construction market of the People's Republic of China (PRC). *Construction Management and Economics*, 27(11), 1141–52. This paper describes research into the emergence of a highly developed form of design build in China which has many of the characteristics of a total construction service. It provides interesting advice for customers on the selection of competent companies.

Chapter Ten
Implications for Industry

10.1 Introduction

This chapter describes the practical implications of the theory of construction management for customers and construction companies. The most complete practical application of the theory which is also the approach that delivers the highest levels of efficiency is a total construction service as described in Chapter 9. The other major approaches currently used in practice which are described in Chapters 5 to 8 include advice for those undertaking key roles in projects using these approaches. However, a major implication of the theory of construction management is these other approaches should be regarded as significant steps towards the greater efficiency provided by a total construction service. Therefore this chapter provides practical encouragement for customers and construction companies to adopt this most efficient approach for all their construction projects and programmes.

It is acknowledged that many experienced customers and construction companies have preferred ways of working which they have developed over many construction projects. However, even these experienced practitioners may benefit from the systematic application of the theory of construction management which underpins this chapter. It may help them avoid the common trap of forcing inappropriate construction projects into their preferred methods. It is understandable to want to use a familiar approach but it may not always produce the best possible outcomes. It is worth considering alternatives because construction projects are never absolutely identical, technologies develop, market conditions change, and the performance of even the most professional of construction people may become jaded. Therefore, all customers and construction companies should review their preferred methods at regular intervals. They may be guided in doing so by this chapter which describes the practical implications of the theory of construction management for those making key decisions about construction projects.

10.2 Implications for customers

Customers have a number of key decisions to make if they are to initiate a successful construction project. These actions are described in this section and illustrated in Figure 10.1.

Construction Management Strategies: A Theory of Construction Management,
First Edition. Milan Radosavljevic and John Bennett.
© 2012 John Wiley & Sons, Ltd. Published 2012 by John Wiley & Sons, Ltd.

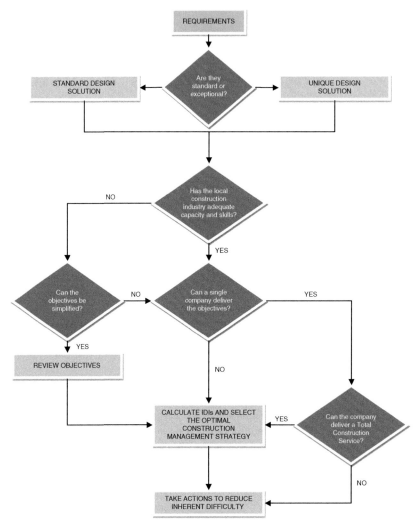

Figure 10.1 Customer's Key Decision Chart.

Ideally customers have a choice between companies offering competing total construction services. Some companies will specialise in basic facilities of reliably good quality which meet most of the customers' needs and can be delivered quickly at a low price. Others will offer more sophisticated, high-quality facilities which meet all the customer's needs delivered at a predetermined date agreed with the customer at a fixed price which provides good value. Still others will provide individually designed, high-quality facilities of obvious aesthetic merit which meet the customer's needs in original ways delivered at a predetermined date agreed with the customer at a fixed price that provides good value for the individual customer.

In this ideal situation customers will be guided through a set of decisions about their new facility which will clearly identify the best buy. They then enter into a simple, clearly worded contract with a construction company; their new facility will be completed at the agreed time and date; their organization will be trained in using the new facility which provides everything the customer could possibly expect; and finally they pay the agreed price.

The ideal scenario has wide implications for how customers are involved in the construction process, particularly with the advent of BIM-supported digital technologies. The experience of procuring a new facility has many similarities to purchasing a sophisticated vehicle. While a customer has a wide selection of overall designs, the majority of components are standardised to enable highly-efficient prefabrication. A customer's involvement in the design is therefore limited to agreeing the overall appearance, the organization of spaces and the selection of standard components. The result is a unique design but it results from a highly organized and standardised system which customers can trust to deliver good value.

In most countries construction practice is far from ideal and customers are forced to play decisive roles in determining many details of the design and also selecting and being involved in the construction management approach used on their projects. This is why it is in the customer's immediate best interests and it is certainly in the best interests of the construction industry and future customers if they aim at using a total construction service. However, in many situations customers find they can do no more than move the construction industry a little closer to towards the ideal of providing a total construction service.

Independent advice

The first task for customers when the local construction industry does not offer a total construction service is to decide whether they have the knowledge and experience within their own organization to make the key decisions in respect of their new facility and deal with the project organization they entrust with the necessary construction activities. Customers who do not have these competences should seek the advice of a project management company experienced both in the type of facility they need and the local construction industry. Initially at least the project management company should be employed to provide strategic advice with no expectation they will be involved in subsequent stages. They may well subsequently be retained to ensure the customer's organization does everything required to ensure the new facility is produced efficiently and meets all the customer's needs, but initially the customer needs independent advice.

Customers who have the required knowledge, experience and leadership skills inside their own organization will normally appoint an internal project manager to be responsible for ensuring the new facility is produced efficiently and meets all their needs.

In practice many customers appoint a designer to guide them through the initial stages. This approach tends to ensure the design is very thoroughly considered but often at the expense of other aspects of the customer's overall objectives. This can lead to an unsuccessful project because the initial stages of construction projects require a broad view of aesthetics, function, quality, time, cost and probably many other issues important to the customer.

Customers wanting a technological complex facility may seek advice from several independent designers and specialists. At least some of the advice may be based on inadequate knowledge of what works in reality leading to suboptimal solutions being accepted by the customer. It is not uncommon in such situations for the final details to be decided only when construction is well underway causing a substantial amount of waste and rework. The most likely outcomes are a less than optimal approach being adopted leaving the customer with an unnecessarily expensive and unsatisfactory facility.

In all these situations, whatever range of design and technical advice is needed, it is sensible for the customer to be guided by an independent project management company. It is important the company is experienced in the particular type of facility required but given that, they are most likely to ensure the customer makes well-informed decisions.

Briefing

The customer's first major responsibility is to ensure the brief fully describes the new facility they need and identifies all the constraints which must be observed in undertaking the necessary construction. Efficient construction is more likely if the brief is produced as early as possible and is not changed during the resulting project.

This primary responsibility does not in any way imply the brief should be rushed. Rather it means the brief should be fully and carefully considered and expert advice sought on aspects of the new facility not within the customer's competence. All the individuals and organizations likely to be affected by the new facility should be encouraged to discuss what is proposed and comment on decisions which they regard as important. This close involvement is sensible whatever approach is used to ensure the new facility meets all the customer's needs as fully and effectively as possible.

The key objectives of the project should be expressed in explicit terms which enable the project organization's performance to be measured. The first draft of the agreed objectives should result from the brief. It will be developed as more members of the project teams are selected and their interests are incorporated in the agreed objectives. In this way all subsequent stages of the project are guided by a set of agreed objectives which are understood and supported by everyone involved.

An important decision which very directly affects the agreed objectives is whether the new facility can use an established design or the customer's particular needs require an original design, individual details or innovative technologies. There are many possible reasons for deciding not to use an established design. A customer may need distinctive spaces, want to make a powerful statement about the style or status of their organization, have a constricted site or need to alter an existing building. Whatever the reasons, a decision that no established design is suitable should be taken carefully because prototypes always take far more time and are considerably more expensive than established answers. It therefore makes sense to use an established answer if one exists which reasonably meets the customer's essential requirements. The existence of good established answers and a mature understanding of their benefits are not well established in construction because much of the industry implicitly assumes each project is unique. As a result construction companies in many locations are not well prepared to use established designs efficiently. There are exceptions to this generally inefficient approach and the range of situations found in practice give rise to customers' second major responsibility.

Construction management strategy

The customer needs to ensure the most appropriate construction management strategy is used. Although the ideal is to use a total construction service, if this is not well developed in the local construction industry, customers should ensure the choice of strategy is considered very carefully. Many customers will need a range of expert advice to guide them through this key stage.

A good starting point for the customer is to insist all their requirements are described in a simple and clear contract with a single construction company which takes total responsibility for delivering the new facility. This serves to focus their advisors' attention on the benefits of a total construction service. However, it may become clear this strategy is not possible in the context of the customer's requirements and the capabilities of the local construction industry. In these circumstances it is worth considering using an established company from outside the local area. This inevitably involves risks because the foreign company may be surprised by local conditions, regulations, culture or other characteristics which influence construction actions. Nevertheless the customer's requirements for new facilities may be large enough to justify working with an external company.

Many customers faced with inadequacies in the local construction industry may still decide their best interests are served by working with local companies. In this situation, a key decision for customers is whether they want one construction company to be responsible for constructing their new facility or whether they want to divide this responsibility between several companies.

In making this decision the customer should consider the extent to which they want to be involved in the project. Employing one construction company means responsibility for the new facility is clear. However, it may be difficult for a customer to influence decisions taken inside the single organization, particularly if it is not experienced in providing a total construction service. Therefore in deciding to entrust their project to a single company, customers should check there are effective arrangements in place which allow them to influence the project whenever they decide this is necessary.

Employing multiple construction companies gives the customer several points of view that may help ensure good decisions. It does mean the customer will be closely involved in the project which can be time-consuming. It also means the customer may be faced with disputes and claims for additional time or money because responsibility for the new facility is unclear.

The choice of a single company or multiple construction companies has a direct influence on the choice of construction management strategy. Whatever the choice, customers should ensure a range of construction management strategies are considered. The essential point of this is to understand the practical implications of their requirements on the costs and benefits. It may be some of their requirements in the context of the local construction industry impose costs which exceed any likely benefit. Customers should ensure they are aware of any such traps as they agree the brief.

It maybe for example that demands for high quality, fast construction and very safe construction ask too much of the local construction industry. The potential outcomes could include high costs, variable quality, delays and a high number of accidents. Aiming at more modest objectives may produce lower costs, better quality, and earlier completion, and reduce the likelihood of accidents. Customers should ensure these issues are considered taking account of the local construction industry's capabilities and the potential of the local environment to interfere with the project. An effective way of doing this is for customers to ask their advisors to calculate the Inherent Difficulty Indicators for all the construction management strategies being considered. Indeed customers may wish to go further and ask their advisors to undertake detailed simulations of the likely performance of the alternative strategies. These early calculations and simulations need data based on realistic assumptions about the performance of the construction companies that will be selected to undertake all the subsequent stages of the project. Chapter 11 describes the research needed to ensure this can be achieved.

Even customers with highly-developed construction strategies can be caught out if they take their approach into an alien environment. Their construction may be subjected to interference of a nature and intensity which makes their approach inappropriate. The construction site may provide conditions which are outside their experience and cannot be handled efficiently by their approach. Where their approach depends on local support or competence, the project may be in a locality where this is not available on any acceptable basis. It is sensible therefore to review even the most highly developed approach in terms of the theory of construction management to ensure it fits all the circumstances of a project before deciding the strategy.

The choice of a construction management strategy is essentially a process of balancing costs and benefits to produce the maximum efficiency. It is an iterative process which begins by simulating the proposed construction actions as they would be undertaken using the local construction industry. This means calculating all the outcomes likely to feature in the agreed objectives if the established local construction management strategy is applied. This establishes the inherent difficulty which provides a useful datum in considering alternative construction management strategies.

The next step in selecting a construction management strategy recognises that each of the actions indentified by the theory of construction management reduces the inherent difficulty of construction. The actions may involve additional costs in providing managers, requiring more experienced construction teams, equipping them with more sophisticated machines and equipment, allowing them more time, protecting construction from interference, investing in training, better safety and welfare provisions, and other initiatives aimed at improving efficiency. At this stage the Inherent Difficulty Indicators provide a reliable quick means of evaluating the options.

When one or two strategies emerge as potentially the most effective, the benefits should be calculated by simulating the effects of selected sets of actions. This means calculating all the outcomes likely to feature in the agreed objectives. The benefits are the differences between the outcomes if the established local construction management strategy is applied and the outcomes resulting from the alternative strategies. The set of actions which provides the greatest net benefit, taking account of the costs involved in applying the actions, is the most effective construction management strategy.

An essential part of selecting a construction management strategy is to review the agreed objectives. As various sets of actions are simulated, it will become apparent that some objectives have major impacts on efficiency. This should be discussed by the key participants with the aim of agreeing the right balance between objectives, costs and benefits. It is likely the discussions will require further simulations to explore the implications of various sets of agreed objectives and construction management strategies. This iterative process is normally carried out in parallel with the production of the brief and the early stages of the design process because the choice of construction management strategy and agreed objectives may well influence decisions made during these early construction actions.

Procurement

Having agreed a brief and decided on the most effective construction management strategy, customer's need to arrange for the procurement of construction companies to undertake the essential construction actions. In selecting construction companies it is essential to make sure they are motivated to undertake the project to the best of their ability and the construction teams they provide are fully competent and well motivated to undertake the required construction actions.

A key part of the customer's role in the initial stages is to create conditions which encourage everyone involved to cooperate for the benefit of the project. Cooperation provides the most effective basis for efficiency as Chapter 8 describes. Yet many customers think competition is an essential driving force to get the best out of individuals, teams and companies. Competition between organizations can provide powerful incentives to outperform each other. It can be a powerful force in fostering innovation and well used it can guide organizations into searching for ever greater efficiency. However, inside a project organization, competition is destructive and results in individuals, teams and companies seeking their own advantage irrespective of the damage done to their joint activities.

It follows customers need to set up project organizations in which there is every possibility that relationships will be guided by cooperation. When a customer employs a single company to provide a total construction service, they can check the company's track record on similar projects. Provided this demonstrates reliably good performance and satisfied customers, it is usually sufficient to interview the key individuals selected by the company to lead the project and check their understanding of the importance of cooperation.

The customer's role is more demanding when they decide to employ a number of companies to create their new facility. This means, as companies are brought into the project organization, the customer or their project manager should ensure the agreed objectives are fully understood and everyone is fully motivated to achieve them. More than this the rewards provided by the project need to be seen as fair by everyone involved. There should be no temptation or incentive for anyone to concentrate on improving their own rewards at the expense of the overall project. These conditions are essential for cooperation to flourish so companies can achieve high levels of efficiency.

The broad aim of all the procurement decisions should be to ensure the agreed construction management strategy is put into effect. In particular this means ensuring the selected construction companies are capable of providing teams, relationships and performance levels which match the assumptions made in selecting the strategy. For example, a mega-project may be too big for any local construction company even in the most advanced construction markets so any strategy based on a total construction service would be unrealistic. In such a case it would be sensible to consider the experience of partnering and construction management which exists in the local construction companies most likely to be involved in the project before selecting the project strategy. On the other hand, a relatively straightforward office building may well be delivered by a total construction service company and there would be no need to increase the inherent difficulty by selecting any other construction management strategy.

Control

As the project is undertaken, the customer should check that systems are in place to ensure the required actions are undertaken efficiently. This should be an integral part of the established approach of the construction companies and teams but it is sensible for the customer to check that the design and plan fully satisfy the brief and competent feedback and control systems are in place and used properly.

It is equally important to ensure the agreed objectives are discussed regularly so as they develop they continue to be clearly and fully accepted by everyone involved in the project organization. As this process of review and development continues the customer needs to ensure all the requirements of the brief are satisfied.

A key part of these vital controls is feedback on the project organization's performance in terms of each of the agreed objectives. The feedback should provide precise measures of performance in terms understandable to the customer, construction teams, the project organization and companies.

The feedback driven controls need to be applied right through to the commissioning stage during which the customer needs to check that their own organization fully understands how the new facility should be operated and used. The most effective approach to achieving a smooth handover of a new facility is for the customer's staff to work with the companies undertaking the commissioning activities. This greatly increases the chances that the new facility will be fully complete when it is handed over to the customer and they will gain the maximum benefits from using it.

In addition, customers can have a significant influence in fostering cooperation between participating teams as opposed to allowing competitive rivalries to develop. Payment mechanisms are one of the major obstacles in achieving cooperation between teams, particularly during the commissioning stage when a large number of mechanical and electrical teams have to work in congested and limited spaces. In most cases payments follow work in progress and teams are motivated to work in as many available spaces as possible to accumulate more work in progress and therefore receive higher interim payments. While a small number of teams can work in the same space simultaneously, the majority may be tempted to compete for spaces bringing in competitive attitudes and delaying the completion of prematurely started work. It is more effective to base payments on completed work. This helps ensure teams are motivated to complete work and are discouraged from opening multiple work positions. Customers can achieve these more efficient outcomes by demanding that all contracts use payment mechanisms based on paying for completed work and involving themselves in the control systems which ensure work is fully complete before a team leaves any work space. The benefits of such payment mechanisms are twofold. In the first place, micro-control of completed work improves overall quality and reduces the burden of everyone involved at the final handover stage of the project. Second, focusing on completing started work focuses teams on activities which add value and helps avoid time being wasted on competition for the available spaces. This all brings greater stability and certainty to the work environment, reduces conflicts and consequently reduces inherent difficulty.

In undertaking their role in commissioning an individual construction project, customers should particularly keep in mind the construction management propositions concerned with project organizations. These are listed in the box below.

Construction management propositions most relevant to customers commissioning an individual construction project

Construction management aims to:

■ Reduce the number of construction teams involved in a construction project;
■ Improve the quality of relationships between the construction teams involved in a construction project;
■ Reduce the performance variability of the construction teams involved in a construction project;
■ Reduce the external interference experienced by the construction teams involved in a construction project;
■ Select construction teams competent in the technologies required by the project in which they are involved;
■ Ensure construction teams accept the agreed objectives of the project in which they are involved;
■ Ensure construction teams are motivated to achieve the agreed objectives of the project in which they are involved;
■ Foster accurate communication between the construction teams involved in a project organization;
■ Minimise the effort required to achieve accurate communication between the construction teams involved in a project organization;
■ Minimise the length and intensity of negotiations needed to agree the transaction which bring construction teams into a project organization;
■ Ensure construction teams regard the transaction which brought them into a project organization as advantageous to themselves;
■ Minimise the resources construction teams devote to improving the terms of the transaction which brought them into a project organization; and
■ Ensure their construction organization collects, reviews and acts on feedback about the effects of the organization's actions on its objectives.

Construction programmes

Customers that need many new facilities should organize their projects into programmes. This creates the basis for long-term relationships between the customer and the leading construction companies and between the various companies responsible for undertaking a construction programme. Ideally a construction programme should comprise a sufficient number of projects to give all the essential interactions time to develop the characteristics of internal relationships. The customer should actively encourage this particularly by ensuring the benefits are distributed in a manner which all those involved regard as fair.

Various ways of organizing construction programmes have been used successfully based on the following general principles. The projects should be sufficiently similar for them to be within the competence of a consistent group of companies. The facilities can be individually designed but should use a similar set of construction technologies. The start dates for the projects should be spaced so the same construction teams can complete their work on one project and then move to the next one and so on. The basis of all the formal legal and financial issues should be agreed at the outset of the programme so individual projects are not delayed by negotiations over terms and conditions. Starting each new project should be a quick and straightforward administrative process. There need to be provisions covering any default by the customer or the construction companies and equally there should be the possibility of extending the programme arrangement because it is successful.

Many customers, particularly in the public sector, have to invite competitive bids for construction projects. It is possible in many situations to extend these requirements to a construction programme. This is usually achieved by inviting bids for a predetermined set of projects or setting up an arrangement which lasts for a fixed period of time. Private sector customers tend to have more freedom in entering into contracts and so can continue an effective construction programme as long as it continues to deliver benefits.

The primary aim in setting up a construction programme in both the public and private sectors should be to give a competent construction company the confidence to provide a total construction service. Where this is not immediately possible, the aim should be for the customer to work with a company or group of companies to get as close to a total construction service as possible. The aims in both these situations should be to ensure construction fully meets the customer's needs, is comfortably within the established competence of the selected construction companies and teams, and they are motivated to undertake the essential construction actions efficiently. This means making construction as simple and straightforward as possible. It means ensuring wherever possible the selected construction companies and teams have established relationships which ideally have all the characteristics of internal relationships. It means ensuring construction teams have a clear understanding of what they are required to do and have everything needed to complete their work exactly in accordance with the design and plan. It means construction companies investing long-term in research, development and training aimed at improving every aspect of their work taking account of feedback from customers and projects. It means customers are provided with excellent value for the investments in construction and are justifiably delighted with their new facilities.

The essential benefits of construction programmes come from enabling construction companies to standardise their project management approaches across a whole programme. The stability that results from working on a number of projects in a programme may well result in companies being able to participate in future programmes with further multiple beneficial effects. First, programme management can achieve steadily increasing efficiency across the programme because of the feedback loops from project managers that record and report best practice (i.e. what works and what does not work). Second, through continuous learning as a result of the feedback loops the involved teams can continuously identify improvements and achieve far greater standardisation of design, manufacturing, production and commissioning. Furthermore, over time they build established relationships with all the partners in the programme establishing a unique culture optimally suited to a particular customer. This by itself builds a competitive advantage that without the benefit of the construction programme could never be realised due to the continuous changes and lack of stability which characterise one-off projects.

Customers setting up a construction programme should keep in mind all the construction management propositions. Those of most direct relevance are listed in the box below.

Construction management aims to:

■ Satisfy the requirements of local or specialised construction markets in ways which ensure their company's long-term survival;
■ Develop construction teams with well-developed skills and knowledge which match the requirements of construction projects;
■ Develop construction teams integrated by established relationships;
■ Improve the quality of support provided to construction teams by the construction companies of which they form part;
■ Foster innovations which match the requirements of construction projects;
■ Ensure their organization uses effective information systems;
■ Establish values which run consistently through every aspect of their organization's work;
■ Ensure all parts of their organization act in ways which are intended and authorised;
■ Ensure their organization's communications are effective and result in common understandings;
■ Ensure their organization's transactions are agreed with the minimum of effort, accepted as fair by all the parties involved, foster established relationships, and are acted on in the spirit in which they were agreed;
■ Ensure their organization forms established relationships with other organizations; and
■ Ensure their construction organization collects, reviews and acts on feedback about the effects of established norms and procedures.

10.3 Implications for construction companies

The most significant implication of the theory of construction management for construction companies is they should think and plan long term. Market conditions in many regions make this difficult. The demand for construction fluctuates, customers commission individual designs, local government agencies impose restrictions and financial institutions impose constraints. Nevertheless, construction efficiency depends on long-term thinking. The challenge facing construction companies is to devise strategies that give them sufficient certainty to invest in the competence and efficiency of their construction teams and at the same time indentify and test new ideas so they can innovate and respond to major changes.

The prevailing short-termism in construction is seen as beneficial by many companies as they believe they can respond to market changes but in fact they leave themselves more exposed to sudden and fundamental changes. Construction companies that focus on the short-term chasing of every new project irrespective of the demands it makes have neither the time or resources to develop core competencies to the level that would give them real market resilience. In essence, short-term focused, project-based companies are more prone to facing difficulties during times of economic adversity than their long-term counterparts. They are continuously involved in different construction management strategies, they change teams regularly and are forced to act as novices rather than being able to enjoy the advantages of being an established major player. Lack of long-term investment in research and development leads them to continuously chase new developments and become 'a jack of all trades, master of none.' When demand falls such companies often collapse like a house of cards because they cannot exploit any strong core competencies or compete effectively in niche markets. Short-term companies inevitably have to rely on a constant stream of particular types of projects, which is equivalent to putting all their eggs in one basket. Only very large short-term-focused companies that operate in numerous niche markets can adequately response to a sudden drop in demand in one market segment. They do this by concentrating on work that comes from other sectors of the market where demand still exists. Such changes are expensive and inefficient but they allow many companies to survive even though profit levels are tiny.

Successful construction companies undertake the construction actions required by local demand efficiently, and steadily and systematically become more and more efficient. This sustained improvement is difficult to achieve. Research into companies which have survived shows they are efficient in undertaking their current work but this is not allowed to become a single minded pursuit of every possible cost reduction. Long-term survival depends on tolerating marginal activities which do not necessarily deliver short-term profits. The benefits of this flexible approach become apparent when market conditions or customers' demands change because the best response is often found in a small dedicated group of enthusiasts who have been allowed to develop a previously marginal activity. In contrast, companies that focus exclusively on their mainstream activities are more likely to be forced out of business when faced with major changes.

In the long term successful construction companies develop attractive, standardised facilities or at least become competent at actions which make a significant contribution to attractive new facilities, support their physical products with sophisticated customer services, and market the resulting packages of products and services. In addition the company needs to remain flexible enough to respond to changes in demand. Indeed in the long term, construction companies should lead market changes by developing attractive new total construction services and marketing them to potential customers.

The construction management propositions which are relevant to all construction companies are listed in the box below.

Construction management aims to:

- Satisfy the requirements of local or specialised construction markets in ways which ensure their company's long term survival;
- Develop construction teams with well developed skills and knowledge which match the requirements of construction projects;
- Develop construction teams integrated by established relationships;
- Improve the quality of support provided to construction teams by the construction companies of which they form part;

- Foster innovations which match the requirements of construction projects; and
- Ensure their construction organization collects, reviews and acts on feedback about the effects of established norms and procedures.

10.4 Construction company strategies

The crucial strategic decision for construction companies is whether they will provide a total construction service so they are responsible for producing complete facilities or specialise in activities which contribute to the production of new facilities so their responsibilities are limited to their own actions and the effects of those actions.

Closely related to this crucial strategic decision is whether they have direct contracts with customers or act as subcontractors to other construction companies. Those companies that provide a total construction service have direct contracts with customers. The situation for specialist companies is much more varied. Depending on their role within individual projects they may have direct contracts with the customer which is most likely to be one of a number of linked relationships with the customer or they may be employed by other construction companies which means from the customer's point of view they are subcontractors.

Subcontractors are often forced to accept the greatest risks but even these companies can achieve greater stability through standardisation and prefabrication. In effect, subcontractors can avoid a lot of uncertainty and risks by developing standard solutions and components which transform them into suppliers in a true manufacturing sense rather than subcontractors forced to bear the brunt of construction's uncertainties. Standard components have another great benefit. They can be offered to the market as stand-alone solutions to designers' problems irrespective of the company's involvement in the particular project. In addition, prefabrication requires research and development which enhances the company's core competencies. At the same time it helps the company avoid much external interference and so minimises the inherent difficulty they face.

These strategic decisions influence the character of the work the company does and the relationships they need to develop. The next sections describe the implications for construction companies which decide to provide a total construction service. The implications for some types of companies which specialise more narrowly are discussed later in the chapter.

10.5 Implications for construction companies providing a total construction service

A total construction service may be based on a narrowly specialised range of facilities. Practical examples include housing, commercial offices, warehouses, hospitals, student accommodation and chemical processing plants. This is well illustrated by the HUF Haus GmbH case study in Chapter 9. Some companies aim to be competent at a number of distinct types of facilities so they can respond to changes in demand. A few aim to be able to tackle a wide range of facilities or indeed to be able to undertake any construction project.

A wide competence almost inevitably means the company needs to be very large. It takes decades to build up the size needed to enable a company to meet the demands of every potential customer. There are risks in being small and too specialised and different risks in being big and widely competent. The key to surviving in all these situations is marketing the company's specific products and services in ways which attract customers.

A great risk faced by very large companies is that they degenerate into an internal world in which departments, divisions and other sections within the organization are divided by internal boundary relationships. The central problem is not the overall company size. It is much more an issue of finding the right size of constituting departments, divisions and other sub-units and carefully building effective relationships between them all. Many large organizations outside construction now recognise sub-units should be relatively small and everyone involved needs to understand their activities are interdependent and they must avoid creating internal boundaries. Whether the optimum number of employees per sub-unit is 100 or more depends primarily on each company's internal culture and way of work but as a general rule, large sub-units lead to internal as well as external isolation.

The problems are often most acute in construction companies which have resorted to mergers and acquisitions in the hope this will strengthen their core competencies and avoid contractually induced boundaries. All too often mergers and acquisitions do nothing to eliminate boundaries between culturally diverse organizations and can all too easily create imbalances between teams that need to work together with consequences which are significantly worse than any resulting from contractually-induced boundaries. It is of paramount importance to understand that effective internal relationships do not result from administrative actions. They develop over time with careful attention given by everyone involved to effective communication, developing a common culture and standardising on products and processes found to be excellent.

Marketing

Construction generally tends to be weak at marketing. It is often thought that effective marketing depends on expensive advertising campaigns. They may be appropriate for a few construction companies but other aspects of marketing are far more relevant for most companies. The essential basis for effective marketing is a deep understanding of potential customers' needs and interests. This is built up by working with potential customers to identify ways in which new facilities can help their organization. It is worth considering all aspects of the use, management and financing of the customer's existing facilities in looking for ways they could deliver more benefits and greater profits.

This market research is essential preparation for construction companies to develop products which give customers real choices in how their key interests can be satisfied. Once the products are available they need to be marketed by three distinct kinds of specialists.

The first element of marketing is provided by experts in the use of facilities. They need to show customers and their organizations how a new facility could help them. This should be backed up by clear descriptions of the options the company is able to offer. Effective approaches include visiting existing facilities or using virtual reality displays of possible facilities and discussing how they relate to the customer's business. Workshops which bring together represen- tatives of users, creative designers, experienced managers and researchers into aspects of the particular type of facility can help build a deeper understanding of the potential benefits. The aim is get potential customers excited about the idea of commissioning a new facility. For instance, HUF Haus GmbH use a number of show houses all over the world that are used for demonstration purposes. These are fully equipped to represent a finished product so potential customers can see what a real-life house will look like and how it is different from the competition. They also have a so called 'HUF village' in Germany where a number of different designs are presented in a real-life situation. Similar approaches could be taken for other types facilities but normally, once a new facility is built and occupied, it is rare for anyone to know who designed and built it.

The second element of marketing is provided by experts in facilities management. They work with the customer's staff responsible for running and maintaining their organization's facilities to help them understand how a new facility could help them work more effectively and efficiently. The benefits may come from sophisticated controls which give the facilities managers more options in running the facility: a layout which makes the facility inherently easier to use or other new technologies which make the facility inherently more efficient. Any of these improvements may deliver significantly lower and more sustainable life-cycle costs. As these issues are discussed, the experts will describe the range of aftercare services they can supply and explain how this can make the work of facilities mangers more effective. An important example of what can be achieved is provided by HUF Haus GmbH who have established HUF Clubs, which are exclusively available to owners of HUF houses in a particular region. Their designs for individual houses have established a cult status and some owners have started building loyalty clubs. HUF regularly organizes events and maintains contact with the owners of their houses. The purpose of this approach is at least twofold. The first is that it provides postoccupancy evaluations through which HUF can improve their products based on feedback from customers. Even more importantly is the word of mouth marketing which has established HUF at the pinnacle of the German housing sector. All this depends absolutely on the company following its core principles of ultimate quality and unique design aimed at fully satisfying every customer's needs and wherever possible exceeding them.

The third element of marketing is provided by experts in finance. They need to be competent to deal with the interests and concerns of the financial depart- ments of major international companies just as effectively as the simpler needs of individuals needing a new facility. A key part of this expert advice is based on understanding the income and other financial benefits which can flow from new facilities. This deep knowledge provides the basis for proposing financial packages which enable potential customers to afford to invest in a new facility.

In all of these crucial marketing activities it is vital to identify the key decision makers in the customer's organization. These are the individuals who will decide to invest in a new facility or stick with what they already have, and will choose between the construction company's products and those of a competitor.

A further important element of marketing is feedback from customers both during projects and subsequently when their organization is using the new facility. Inevitably the construction company will be all too aware of any problems since most customers complain if their new facility is in any way disappointing. A joint discussion of the facility between the customer and key members of the company's staff involved in the project can identify lessons for all parties. Customers may learn how to use their facility more effectively and the construction company may get a better understanding of the customer's real needs.

The aim of the various marketing activities is to persuade customers to order a new facility from the company. It is therefore vital that the company can deliver everything discussed with potential customers. In other words marketing must be firmly based on the company's capabilities. It also plays an important role in identifying ideas and innovations which would benefit the company. This means there need to be close relationships between the marketing specialists and the rest of the company. In directing these important activities, the construction management propositions provide important guidance. The most relevant propositions are listed in the box below.

Construction management aims to:

- Satisfy the requirements of local or specialised construction markets in ways which ensure their company's long-term survival;
- Develop construction teams with well-developed skills and knowledge which match the requirements of construction projects;
- Develop construction teams integrated by established relationships;
- Improve the quality of support provided to construction teams by the construction companies of which they form part;
- Foster innovations which match the requirements of construction projects;
- Ensure their organization uses effective information systems;
- Establish values which run consistently through every aspect of their organization's work;
- Ensure all parts of their organization act in ways which are intended and authorised;
- Ensure their organization's communications are effective and result in common understandings;
- Ensure their organization's transactions are agreed with the minimum of effort, accepted as fair by all the parties involved, foster established relationships, and are acted on in the spirit in which they were agreed;
- Ensure their organization forms established relationships with other organizations;
- Ensure their construction organization collects, reviews and acts on feedback about the effects of the organization's actions on its objectives; and
- Ensure their construction organization collects, reviews and acts on feedback about the effects of established norms and procedures.

Company work characteristics

Having won an order for a new facility, the company needs to assemble the core team of the project organization. This is the team that will work with the customer to agree the brief, and the terms and conditions on which the company will undertake all the necessary work and then ensure the new facility is constructed efficiently and fully satisfies the customer. The core team is normally led by a project manager who provides the primary point of contact between the customer and the company throughout the project.

The core team should include individuals experienced in all aspects of the construction of the type of facility the customer needs. It is good practice to call on other points of view during the initial discussions with the customer so real options can be considered before making firm decisions. Ideally all the individuals involved at this stage are employees of the construction company but where this is not the case, they should be employed as specialist consultants. These arrangements should not be of any concern to the customer who should feel they are dealing with a single integrated organization.

It is good practice during these early stages to evaluate the physical, political, economic and social environments which may influence the project. The company needs to understand the potential for external interference and consider how any serious threats can be avoided or mitigated. This risk management should continue throughout the project to give the construction teams the best chance of completing their work as planned. For instance, although the company itself can be viewed as traditionally German, HUF Haus GmbH operates all over the world. They are responsible for submitting planning applications on behalf of their customers and they ensure their documents far exceed the local basic requirements thus raising the stakes for their competitors. They understand that every new market is a potential legislative minefield and are determined to avoid problems by exceeding the expectations of relevant authorities regardless of the initial costs. They are highly focussed on establishing their presence in the market and are aware that this comes at a price but in the long term the benefits far exceed the initial efforts and costs.

As the project progresses, the core team will involve other construction teams to undertake specialised activities. These may come from inside the company or be subcontractors. As with the core team, these arrangements should in no way diminish the integrity of the project organization.

The company's aims should be for all construction teams to have clearly defined tasks which are comfortably within their competence. They should have fixed start and completion dates and minimal interaction with other teams. The working arrangements must include quality and safety controls administered by the team and regularly monitored by independent inspectors who maintain detailed records of their tests and inspections.

Achieving all this means the company needs to invest in developing its range of skills and knowledge so it can undertake a growing range of projects with ever greater confidence and efficiency. The theory of construction management provides directly relevant guidance in making these investment decisions, the direct outcomes of which are training, and research and development.

Training deals with the technology of individual team's work. It also helps them develop as a closely integrated team so they work together ever more effectively. It broadens the team's range of skills so they can increasingly manage their own work without the need for specialised managers. In particular it provides training in quality and safety control. It helps teams systematically identify better ways of working and make suggestions for improving the facilities the company constructs.

A construction company's investments in research and development usually begin on a small scale by working with external research institutes. This may be in providing test sites for new components or access to projects to develop improved techniques. As companies learn the value of research and development, they are likely to suggest new ideas and arrange for staff to spend time in high-quality research institutes. Ultimately the most successful companies have their own research institutes which ensure all parts of the company have access to the most robust current ideas and technologies.

Company relationships

The quality of relationships between construction companies and individual teams has a direct influence on efficiency. Ideally good relationships are an integral part of every company's work. This ideal has to be systematically developed and sustained.

The starting point is developing highly-effective relationships inside the company. All the distinct departments and individual construction teams should automatically behave on the basis of efficient internal relationships in dealing with each other. Senior management should actively monitor and foster this level of efficiency. Problems should be dealt with as they occur. Ideally good solutions are agreed by the individuals, teams or departments directly involved. However, if a problem is not solved quickly in this way, it should be referred to more senior managers who should have a short time to find an answer. If that does not work, it should be passed to top-level managers to decide on a solution and ensure it is implemented. Efficiency is absolutely dependent on not allowing internal problems to fester. Weekly design-execution review meetings can be effective in identifying and eliminating recurring problems and technical difficulties with the aim of continuously improving products, particularly where a large proportion of the work is standardised and requires prefabricated components to be assembled on site. The review meetings serve as a platform for designers to meet production teams to discuss any design-induced difficulties. This ensures standard design solutions are continuously improved for the benefit of customers and the total construction service company. In this way new developments are not limited to the outputs of research and development teams but stem from the everyday work of teams on site and their interactions with designers.

Most construction companies need to work with other companies. Even companies providing a total construction service do not always have all the required skills and knowledge available internally. It is a huge advantage to everyone involved if relationships between companies provide all the advantages of internal relationships. Achieving this takes time and effort at a senior level in both companies involved in any given relationship. It is good practice for a senior manager in each company to be given an explicit responsibility for ensuring high-quality relationships develop. The relationship managers need to be actively involved, meet regularly to review any problems and identify ways of fostering ever better relationships. They need to be closely in touch with their own staff and it often helps for them to lead joint meetings between key specialist managers in both companies.

The aim is for the companies to establish effective relationships at all levels. These need in particular to influence relationships between the construction teams involved in any one project. This is a key part of achieving the ultimate aim of effective construction companies which is to enable construction actions to be undertaken efficiently. The box opposite lists the construction management propositions of greatest relevance to construction companies' work and relationships.

Construction management aims to:

- Reduce the number of construction teams involved in a construction project;
- Improve the quality of relationships between the construction teams involved in a construction project;
- Reduce the performance variability of the construction teams involved in a construction project;
- Reduce the external interference experienced by the construction teams involved in a construction project;
- Select construction teams competent in the technologies required by the project in which they are involved;
- Ensure construction teams accept the agreed objectives of the project in which they are involved;
- Ensure construction teams are motivated to achieve the agreed objectives of the project in which they are involved;
- Foster accurate communication between the construction teams involved in a project organization;
- Minimise the effort required to achieve accurate communication between the construction teams involved in a project organization;
- Minimise the length and intensity of negotiations needed to agree the transaction which bring construction teams into a project organization;
- Ensure construction teams regard the transaction which brought them into a project organization as advantageous to themselves;
- Minimise the resources construction teams devote to improving the terms of the transaction which brought them into a project organization;
- Ensure their construction organization collects, reviews and acts on feedback about the effects of the organization's actions on its objectives.

Reputation

Construction companies should actively monitor and develop their reputation. The foundation for this is successful projects and satisfied customers. It is however, necessary to ensure the company's achievements are widely known. This means: building good links with the media; providing news of exciting designs, new technologies and innovative solutions to difficult construction problems; and be readily available to comment on construction issues. HUF Haus GmbH for instance achieved a major breakthrough in the United Kingdom housing market when their houses were featured on a popular property television show. This is but one example of the ways in which they build their reputation. Their website contains press releases from all over the world and they maintain close links with the local media in every region where they operate. Only a handful of construction companies manage to achieve similar exposure on a global scale but in most cases this is driven by the media and the exceptional nature of individual projects. All too many construction companies are reluctant to accept media exposure.

The benefits of a good reputation are considerable. It is easier to attract serious customers and reach mutually beneficial agreements with them. It is easier to persuade other construction companies to cooperate on complex projects or programmes. It is easier to get the support of local communities and powerful individuals in those communities. It is easier to attract and keep good staff. All these things are easier to do for construction companies that success-fully provide a total construction service. However, for many reasons the majority of construction companies are more narrowly specialised and the following sections describe major implications of the theory of construction management for some of the more significant of these specialist companies.

10.6 Implications for project management companies

It is important to agree the responsibilities and authority of a project manage-ment company. This applies to all specialist companies but is a particular issue with project management companies because their responsibility is normally to carry out specified duties designed to ensure the project is a success but they do not guarantee the project outcomes. In other words if the project manage-ment company undertakes their duties in accordance with established best practice but nevertheless the project is completed late or its costs escalate or it fails to function as expected, the project management company is not liable for the customer's loses. Ensuring these issues are clearly understood is vital in establishing effective relationships with the customer and other construction companies involved in a project.

   The normal role for a project management company is to guide the customer through the decisions they need to make and undertake the actions required of the customer on their behalf. Thus the sections earlier in this chapter which describe the implications of the theory of construction management for customers cover the issues which affect project management companies.

10.7 Implications for construction management companies

The normal role for a construction management company on an individual project is to work with the customer and designers in producing the brief, design and plan for a project and ensure they are put into effect. As Chapter 7 describes, construction management companies are particularly responsible for producing the plan and undertaking procurement. Construction manage-ment companies have undertaken these roles on many of the largest and most challenging construction projects over the past three decades. The best of these companies already act on the basis of the construction management propositions as the case study in the box below demonstrates.

> This case study describes how an outstanding construction management company decided to invest in providing training for the suppliers and contractors they work with in undertaking construction projects and pro-grammes. Mace recognised they had reached a stage where further improvements in efficiency depended on improving the management processes of their industry partners. This led them to set up the Mace Business School to tackle this challenging task. The significance of this decision is based on Mace's short but impressive history.

Mace was established in 1990 by a group of five people determined to create an outstanding construction management company in the United Kingdom. They have been highly successful as Mace now employs over 2,800 people worldwide and the company is responsible for many of the most iconic construction projects of the past two decades. These include the London Eye, London's City Hall, Heathrow Terminal 5, the Venetian in Macau, which is the world's largest casino resort, the British Museum Grand Court and construction for the Manchester Commonwealth Games. Their most prestigious achievement to date is being one of the three companies forming the CLM consortium which is project managing the construction of the venues and infrastructure for the London 2012 Olympics. This firmly establishes Mace as a premier construction company. Perhaps as a reflection of this, they are constructing the tallest building in Europe, the Shard London Bridge which will be 310 metres high, designed by Renzo Piano for the Sellar Properties Group.

Mace's rapid development has inevitably been influenced by major changes in economic conditions which create new opportunities and demands from customers. A guiding principle in responding to these external forces has been to maintain and develop their outstanding management skills while finding new ways of applying them. As a result the company is now organized into three broad divisions. One provides consultancy services which range from developing a customer's business case, through all stages of construction to managing the new facility. A second undertakes construction projects and programmes on a construction management or fixed-price basis. Although initially Mace was primarily involved in large, private sector developments, it now also operates in the public sector. Indeed it has found innovative ways of responding to the needs of major public bodies and has responsibility for the construction programmes of several local authorities including Hertfordshire County Council. The same strategy of meeting the real needs of major customers resulted in Mace's third broad division. It runs their international business and typically is involved in about 30 countries worldwide. This global expansion was greatly helped by early work in the United Kingdom for the Commonwealth Office which owns buildings in most countries. Mace also has a number of specialised companies providing services which include facilities management, residential construction, cost consultancy, contracting and the Mace Business School.

The Mace Business School was established to help improve the performance of the construction companies which undertake the manufacturing, production and commissioning on Mace projects and programmes. The need for this emerged as Mace's project managers found improvements in their overall performance being inhibited by failings on the part of their partner construction companies. These rarely related directly to construction actions but concerned the companies' management inputs.

Mace's approach is based on establishing agreed objectives for their construction projects and programmes which reflect the customer's needs and are expressed in measurable terms. It is often vital to be able to demonstrate the agreed objectives have been met or exceeded. A key to driving this essential control is for construction companies supplying construction teams to projects and programmes to provide full and accurate information on time. Even the best of the construction companies which Mace works with failed to consistently meet these tough but essential requirements. As a result plans were compromised, for example, by companies providing method statements late or incomplete. Asking for revised method statements often made things worse as hastily prepared changes introduce new problems. This pattern too often continued throughout subsequent construction stages as, for example, essential health and safety information was not produced accurately or on time.

The persistence of the problems was in some ways surprising since Mace has a sophisticated system for identifying the best construction companies to work with. Their system is based on the measured performance of the partner companies on Mace projects and programmes. It provides regular checks on all of the 160 preferred companies and 200 approved companies. Despite this rigorous system, the failings were such that some of Mace's managers started informal workshops to explain to the construction teams on their projects what information they were contractually required to produce. This provided benefits but the lessons were not carried back into the companies and so each new project faced the same problems. This showed that informal, free training was neither effective nor viable but finding a robust answer was difficult because of the scale of Mace's operations.

The Mace Business School was established following an extensive series of interviews and discussions with Mace's project managers to understand the learning needs. The school initially concentrated on 20 key companies who were offered the new status of recommended contractors once their managers had completed the Mace training. The companies were selected on the basis of their impact on Mace projects and programmes. The factors taken into account included: the company's impact on costs; their provision of expensive machines; being on site for long periods e.g. scaffolding companies; the need for unique skills and knowledge; or their work was on the critical path and so could disrupt overall progress.

Training takes place on ten individual days and covers subjects which include: commercial management, construction management, environmental impact, health and safety, document management, procurement, trade contractor coordination and logistics. There are also specialised days for services contractors which cover: coordination drawings, quality management and commissioning. Individuals keep a record of their training which is stamped as they complete each day's training. When they have successfully completed all ten days they are issued with a passport which is similar in form to a credit card and confirms their status as a member of a Mace recommended company. The passport is valid for one year and is revalidated annually. The validation checks current training against that received by the passport holder and identifies the need for any updating. There are now over 1,500 Mace passport holders.

An important initiative by the business school came from the recognition that a high proportion of the construction managers they were training had no formal qualifications. They had initially worked as tradesmen and gradually been promoted to more and more responsible positions. More than half only had the training needed to meet specific statutory requirements. These people were technically competent and could often see the best answers to problems but were unable to support their ideas with convincing and reasoned arguments. This inhibited them in dealing with qualified professionals such as architects, engineers and surveyors.

The business school is using the Chartered Institute of Building's Experienced Practitioner Assessed Programme to give experienced managers an entry route to one of the major United Kingdom professional institutes responsible for construction. This is being achieved through a course which covers the required four modules: construction technology; health, safety and the environment; contract and commercial practice; and management. The first two modules are based on assignments and the other two on formal examinations. The final stage is a formal professional interview. The whole course is spread throughout one year. The business school is justifiably proud of the fact that all the first cohort of Mace students achieved a distinction and two won medals for their outstanding performance.

Throughout its various initiatives the business school recognises that learning is a two-way process. In explicit recognition of this, the school creates opportunities for groups of directors from the recommended companies to talk about their joint business. These executive briefings serve to develop common understandings about business strategies. They reinforce an understanding that opportunities always exist but there is a skill in recognising them and deciding which to pursue. New developments involve risks and so it makes sense to base them on established company strengths. Initiatives should aim to develop new customers and enhance the company's reputation as well as ensure long-term, profitable survival. Discussing these issues helps build a common culture shared by Mace and its key partner companies which drives a relentless search for greater efficiency and new business opportunities.

A significant further development by the business school is to run a Master Class for young, high-flying managers from their partner companies. It teaches business as a generic subject. It is not company or construction specific. The Master Class comprises three separate blocks each of three intensive days. It aims to help participants develop the skills and knowledge needed to run a business. At the end of the Master Class the participants make a presentation to Mace Directors. The first Master Class suggested Mace's highly developed management systems are not ideal for smaller projects. The participants acknowledged the systems are highly successful on major projects and programmes but argued smaller projects are different. They concluded their presentation by proposing Mace establish a new business to deal with small, repetitive construction projects and programmes. The Mace Directors thought about this for a few minutes and agreed. As a result Mace has a new business called Project Exchange.

The Business School intends to continue its established courses and continue finding new ways of adding value for all Mace's customers. The company's established measures already show clear improvements in the performance of those partner companies which include managers trained by the business school.

The essential message of this case study is the highest levels of efficiency depend on all construction companies acting to improve their performance and capabilities. Mace cannot deliver world-class standards on its own. It has to work with its partner companies to keep raising their joint efficiency.

10.8 Implications for design companies

Design companies have a decisive influence on the success of construction projects. Their designs directly shape the work of the construction teams responsible for manufacturing, production and commissioning. In many ways their responsibilities (see box p. 253) mirror those of construction management companies except their direct responsibility is to produce the design.

Designers naturally regard the physical appearance and performance of the facility as their primary responsibility. Achieving the best possible designs is a challenging task and can easily lead to issues affecting the efficiency of a project being ignored or regarded as other people's responsibility. This is to underestimate the influence of the design. Every design decision affects costs, value, efficiency and indeed the overall performance of the project organization. To take a very simple example, glass is manufactured in sheets of particular sizes and so it is efficient to design windows which need glass of those sizes. Having to cut glass to fit into windows of a different size or shape is wasteful and expensive. Similar comments can be made about every technology used in construction.

It follows that the most direct implication of the theory of construction management for design companies is they should used well-established technologies in ways which are familiar to local construction companies and their construction teams. This means for each distinct technology designs should require the use of standard components, fixings and junction details. It also means designs should require the use of fixing and junction details between different technologies which are well understood by the selected construction teams. Indeed some of the most effective design companies have well-developed libraries of design details which respect these principles of efficient construction. Many of these successful companies are content to specify the performance and appearance of the major elements of facilities and rely on specialist contractors to select and apply their preferred design details in realising the overall design.

The advent of Building Information Modelling (BIM) and virtual construction (Figure 10.2) creates the possibility for design companies to franchise their signature designs in an almost supermarket style environment. Signature architecture franchising has been proposed by Professor Marck Clayton from Texas A&M University and may well become a major milestone in the way architects sell their services. In essence, virtual construction enables the mass commercialisation of signature architecture based on a single highly acclaimed project. In an architecture supermarket designers would generate significant financial profits by selling regionally specific signature architecture to local construction companies. Design details would be complete and standardised allowing construction companies to use a distinct architectural language to build unique yet standardised buildings where the vast majority of components are prefabricated. This could result in high architectural standards across the board because signature architecture was used on all projects as opposed to being limited to facilities provided for a select few customers willing to pay for a unique one-off product.

Design companies should work on the basis of these principles wherever possible and use non-standard components, fixings and junctions only if this is essential to meet the customer's requirements.

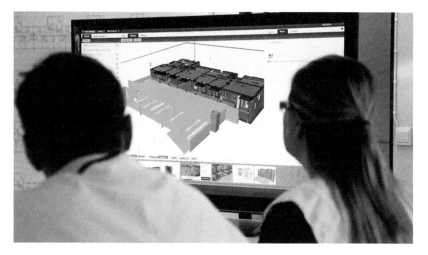

Figure 10.2 Virtual Construction using State-of-the-art BIM Technology. *Source:* Tekla Oy, Building Information Modelling software vendor).

Construction management aims to:

- Reduce the number of construction teams involved in a construction project;
- Reduce the performance variability of the construction teams involved in a construction project;
- Select construction teams competent in the technologies required by the project in which they are involved;
- Ensure construction teams accept the agreed objectives of the project in which they are involved; and
- Ensure their construction organization collects, reviews and acts on feedback about the effects of the organization's actions on its objectives.

10.9 Implications for specialist contractors

The theory of construction management aims to guide project organizations and the companies which form them into creating conditions which enable the construction teams provided by specialist contractors to undertake their manufacturing, production and commissioning activities efficiently. The relevant propositions are listed in the earlier box (see page 247). It follows, if the theory is followed, project work for specialist contractors requires a straightforward application of their well-developed methods. In this ideal situation, specialist contractors can devote time and resources to becoming ever more efficient.

Specialist contractors, like all construction companies aiming to improve their own efficiency, need to think and plan long term. This means investing in the competence of their construction teams. It means ensuring they are supported by effective systems and equipment. It means providing training to ensure they use the best available methods. It means coaching them in cooperative behaviour so they naturally work as real teams and are comfortable establishing effective relationships with other construction teams. It means empowering them to systematically search for ways of improving their performance and the technologies they use. Specialist contractors mostly act as subcontractors. Therefore, they should follow the recommendations described earlier in this chapter which relate to subcontractors. Most importantly this means striving to achieve higher levels of standardisation executed through carefully planned prefabrication. The broad aim should be to achieve the levels of certainty and stability which are common for suppliers in the manufacturing sector.

Specialist contractors should invest in their technologies. This means monitoring feedback, dealing with problems quickly and constantly searching for ways of improving their products. Many good ideas will come from construction teams working on individual projects. Further improvements will be identified by monitoring the performance of their technologies in facilities they helped construct. The various sources of feedback should guide development projects and research in a relentless drive to ensure that every year their technologies are measurably better than the year before. The construction management propositions listed in the box below will help ensure specialist contractors' investments pay off by delivering ever higher levels of efficiency.

Construction management aims to:

- Satisfy the requirements of local or specialised construction markets in ways which ensure their company's long-term survival;
- Develop construction teams with well-developed skills and knowledge which match the requirements of construction projects;
- Develop construction teams integrated by established relationships;
- Improve the quality of support provided to construction teams by the construction companies of which they form part;
- Foster innovations which match the requirements of construction projects;
- Ensure their organization uses effective information systems;
- Establish values which run consistently through every aspect of their organization's work;
- Ensure all parts of their organization act in ways which are intended and authorised;
- Ensure their organization's communications are effective and result in common understandings;
- Ensure their organization's transactions are agreed with the minimum of effort, accepted as fair by all the parties involved, foster established relationships, and are acted on in the spirit in which they were agreed;
- Ensure their organization forms established relationships with other organizations;
- Ensure their construction organization collects, reviews and acts on feedback about the effects of the organization's actions on its objectives; and
- Ensure their construction organization collects, reviews and acts on feedback about the effects of established norms and procedures.

10.10 Implications for other construction companies

This chapter has described the implications of the theory of construction management for some of the key companies in the construction industry. Other types of companies are well advised to use the theory to guide their strategic thinking and day-to-day actions. The descriptions in this chapter provide guidance likely to be relevant to their particular circumstances.

Exercise

Consider the following two examples, and using the customer's key decision chart and propositions in this chapter describe the procedure you would use as a customer to select the most appropriate construction management strategy taking into account the implications for participating construction companies.

Project 1   Tallest building in the world to be built in downtown Abu-Dhabi built according to the latest environmental and technological standards.

Project 2   40-story office building in downtown London built according to the latest environmental and technological standards

In particular, think about the following questions:

1. Is the capacity of the local construction industry adequate to be able to rely on local construction companies?
2. Could you use standard design solutions and what factors should affect the design choice?
3. Looking at the IDI calculations in Chapter 4, what factors would you need to consider when selecting the optimal construction management strategy?
4. What actions could you take to simplify the objectives set out in the brief?
5. What actions could you take to reduce the inherent difficulty of the projects?

Further Reading

The following publications are the source of ideas used in this chapter and provide further information for readers.

The Chartered Institute of Building (2010) *Code of Practice for Project Management for Construction and Development*. Blackwell. The fourth edition of this practical guide provides a wealth of guidance on the effective management of construction projects supported by detailed checklists. The code of practice is particularly suitable for students due to its systematic coverage of all aspects of project management and for practitioners as a quick reference.

Sacks, R., Radosavljevic, M. and Barak, R. (2010) Requirements for building information modelling based lean production management systems for construction. *Automation in Construction*, 19(5), 641–55. The paper proposes a BIM-enabled system to support production and day-to-day control on construction sites using touch screens and other interactive technologies. It defines the requirements and visually depicts a series of user interfaces for such a system on the basis that building information models should be usefully applied in production management to achieve smooth and uninterrupted workflow.

McCabe, S. (2010) *Corporate Strategy in Construction: Understanding Today's Theory and Practice*. Wiley-Blackwell. The book provides a comprehensive overview of corporate strategy and its realisation within the construction context. Its approach of relating mainstream strategic management and the global market environment to the business of construction is of particular importance.

Chapter Eleven
The Future for Construction Management

11.1 Introduction

Unprecedented technological progress in recent decades gives construction management research the tools needed to study the complexity and uncertainty which characterise construction. Building Information Modelling (BIM) is already becoming established at the leading edge of practice. It enables design information to be checked for errors and clashes and evaluated in terms of the likely outcomes in terms of performance, quality, safety, time, cost and other important outcomes.

Total construction companies use BIM to control construction projects through all their stages as a seamless process free of proprietary, confidentiality or interoperability issues. Fragmented projects which rely on multiple companies each with their own sub-models suffer from limited compatibility and rigid organizational boundaries but they still find it worthwhile to use BIM. It is far more difficult and less efficient than within a total construction service company and this efficiency gap is one of the major forces driving a progressive shift towards a total construction service strategy. It is virtually certain as BIM is more widely used it will lead to deeper integration between construction companies that work together frequently.

These developments provide major opportunities for construction management research. Indeed many researchers are already deeply involved in working with companies to explore and foster the use of BIM and related technologies. The full benefits require a new way of thinking about construction and the primary purpose of this book is to provide the essential intellectual basis for the important developments in construction management research made possible by the new technologies.

The fundamental problem facing construction management research is the absence of an established classification scheme for construction projects. This makes it virtually impossible to relate research results to a body of established knowledge and forces researchers to work in isolation from each other. Perhaps more importantly it frustrates and devalues much of the discussion between practitioners and researchers which inevitably leaves construction less efficient than it could and should be.

Practitioners most commonly describe projects in terms of the function of the facility they aim to construct. Thus, they discuss office buildings, hospitals,

Construction Management Strategies: A Theory of Construction Management,
First Edition. Milan Radosavljevic and John Bennett.
© 2012 John Wiley & Sons, Ltd. Published 2012 by John Wiley & Sons, Ltd.

apartment blocks, bridges, roads and so on. These descriptions tell us very little about the construction management task posed by the projects. Practitioners may also describe projects in terms of size and duration. This results in projects being described in such terms as a large hospital project taking 3 years, a minor extension to a primary school taking 6 months or a huge chemical processing plant taking 10 years to complete. The size and duration factors have a more direct influence on the construction management task but the links are weak and inconsistent. The most important construction management research inevitably has to devote considerable efforts to defining a set of construction projects which then form a unique basis for an individual study.

Real progress requires a map of every possible construction project based on the reality of construction management. This will enable practitioners and researchers to relate their work to a wide body of construction management knowledge and experience. Such a map of every possible construction project will allow sets of data from individual projects describing the key concepts in the theory of construction management illustrated in Figure 11.1 to be related to each other. Proposals for establishing a suitable map of every possible construction project are included in the section 'Testing the theory of construction management' later in this chapter.

As Figure 11.1 shows the key concepts in the theory of construction management are practitioners select construction management strategies in response to the inherent difficulty of their projects in order to achieve agreed objectives efficiently. It further shows inherent difficulty depends on the number of interacting teams, the quality of relations between them and performance variability which together determine project complexity and

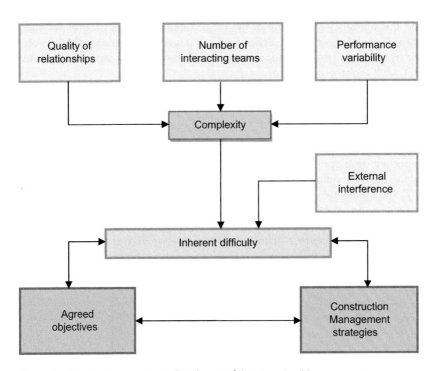

Figure 11.1. Basic Concepts in the Theory of Construction Management.

external interference. An important distinction between these two concepts is complexity results from decisions made by project organizations but external interference is caused by circumstances outside the direct control of project organizations.

This chapter begins by describing the relationship between the key concepts in the theory of construction management and practice. It then reviews the relationship between these concepts and research. These introductory sections provide the basis for identifying the research needed to provide a robust basis for establishing a classification scheme for construction projects. An obvious benefit of this approach is such research will relate directly to the tasks faced by construction managers and so increase the probability of the results being directly relevant to practice.

11.2 The theory of construction management and practice

Chapters 5 to 9 make clear that the three crucial interacting elements influence practice. This evidence of the practical importance of agreed objectives, inherent difficulty and construction management strategies provides the starting point for designing a future direction for construction management research.

Agreed objectives

Construction managers have many different types and levels of objectives. The box below provides a list of subjects which may be included in these objectives. Some are short-term and relate to specific actions. Most objectives are longer-term and relate to sets of linked actions. Beyond this, long-term objectives tend to be concerned with financial investments, security of profits, capital resources, market intelligence, human relations, recruitment, training, innovation, research and similar long-term strategic issues.

- Customer satisfaction with the new facility and the services provided.
- Quality including testing, incidence of defects and rate of improvement.
- Time including predictability and speed.
- Safety in terms of safety provisions, accidents, injuries and deaths.
- Productivity of every team, division and company.
- Costs of resources used at all levels during construction and rate of cost reduction.
- Proportion of time spent on productive work.
- Effectiveness of relationships with suppliers and subcontractors.
- Claims for extra costs or time and rate of reduction in the incidence of claims.
- Life-cycle costs of the new facility.
- Profitability in terms of the size and security of profits.
- Design quality and accuracy.
- Energy use during construction.
- Complaints from neighbours about the new facility and construction actions.
- Materials and components rejected after being delivered to the construction site.

- Efficiency of communications.
- Effectiveness of transactions.
- Staff training, morale and turnover, and speed of dealing with grievances.
- Suggestions for new products and services.
- Development of new technologies and innovative facilities.
- Quality and effectiveness of marketing information.
- Company reputation and market share.
- Effectiveness of company strategies and policies.
- Effectiveness of feedback

Leading construction management practice ensures the individual objectives of those involved in any given set of construction actions are made explicit, discussed and common objectives agreed. A common trap which is avoided in best practice is to concentrate on one or two important objectives and as a result neglect other performance factors. A decision to concentrate on say reducing cost can result in poor quality, late completion and risks being taken with workers' safety.

Ownership is one of the key requirements for the agreed objectives to be delivered satisfactorily. Each objective must have a dedicated owner responsible for its execution and monitoring. In addition, it is imperative that owners are clear what they should achieve so once the objectives are achieved there is no ambiguity about whether they have in fact been achieved satisfactorily. Whether they relate to physical attributes of the facility, such as the quality of individual components, or the behaviour of construction teams, such as the efficiency of communication, the agreed objectives need to be measurable. For instance, tackling a particular problem may involve more than an acceptable number of people and their responses take longer and absorb more resources than would be normally accepted, then although the particular problem is resolved, communication would be judged as inefficient. This failure is likely to be identified only if there is a dedicated owner responsible for the performance of communication chains who has established measurable and agreed objectives for efficient communication. Such an approach demands rigorous tracking of communications which may be viewed as intrusive and bureaucratic, but careful monitoring and response to inefficiencies can lead to far shorter communication chains that benefit every construction team. Depending on the overall size of the project, a dedicated communications officer or team can be responsible for enabling and maintaining efficient communication. Information technology plays an essential role here by enabling the automatic transmission of communicated matter to relevant individuals. In practice, case studies show that resolving even highly unusual and unexpected problems can be achieved considerably quicker through the smart use of information technology and careful monitoring of communications.

The correct approach is to define an acceptable norm in respect of every performance factor and ensure they are all achieved. In addition agreed objectives may require improvements over and above the norm in one or at the most two factors. A mass of evidence from case studies of practice shows that aiming to improve multiple performance factors almost invariably means construction teams fail to meet their objectives.

An important part of agreeing objectives is to agree how performance will be measured. Ideally the construction industry has established measures of performance in respect of all the factors regarded as important by customers and construction companies. Where such measures do not already exist, the companies involved should decide exactly how the achievement of each objective will be measured. Then agreed objectives can be expressed in terms of a precise level of performance for each important factor and as the actions intended to achieve the objectives progress, actual performance can be measured. As the theory of construction management makes clear, this feedback should be produced routinely at regular intervals to give a robust basis for construction management decisions.

Benchmarking

Performance measures are used in a practical approach to improving performance called benchmarking. This management technique grew out of a recognition that many of those involved in leading companies in many industries, including construction, measure their organization's performance. The results are used to establish benchmarks of an organization's normal performance. This routine use of benchmarks is well developed and it led groups of practitioners to use them as a basis for improving their performance. Indeed when organizations routinely measure their performance, it becomes almost irresistible to search for ways of making improvements.

The outcome is a formal process called benchmarking. There are well-researched guides to benchmarking including some which deal directly with construction. They describe how several organizations work together in a cooperative search for ways of improving their individual performance in selected aspects of their work. The guides mainly describe the use of data produced by the organizations but some recognise that various national industries have established key performance indicators which allow individual companies, project organizations and teams to compare their performance with industry norms.

In the future benchmarking can be made more effective by taking inherent difficulty into account. Through the use of IDIs construction companies will be able to evaluate whether some of their projects are outside the norm in terms of inherent difficulty. Some projects may involve more teams than normal with fewer established relationships. There may also be projects that involve teams with greater performance variability making it difficult to achieve an overall objective of minimising inherent difficulty. This may have strategic implications for construction companies as their top management realise that involvement in a particular type of project or with a particular customer carries a high risk of operating in projects which are inherently very difficult.

Benchmarking would become significantly more effective in driving construction performance improvements if organizations could compare their performance with direct competitors throughout the construction industry. This should be a future objective for research which will be facilitated by the proposals described later in this chapter.

Inherent difficulty

The concept of inherent difficulty is not well developed in practice. This is a serious weakness because IDIs would enable companies to develop effective ways of reducing the inherent difficulty of the projects they are involved in. This may be achieved through team consolidation, focusing on uniform performance to reduce performance variability, or by selecting teams with established relationships (internally or externally). Using IDIs is likely to prompt construction companies as well as customers to focus on managing the key complexity factors bringing more stability and higher levels of efficiency into the project environment.

Construction management strategies

Chapters 5 to 9 are organized on the basis of widely accepted categories of construction management strategies used in practice. This gives us:

- traditional construction;
- design build;
- construction management which includes management contracting;
- partnering; and
- total construction service.

There are no theoretically robust definitions of any one of these construction management approaches. There are considerable variations within each of them in the approaches found in practice. As a result construction management research lacks the firm basis which a robust classification of the approaches found in practice would provide.

A large number of case studies of practice describe the effects of construction management strategies which introduce changes in line with the construction management propositions. These results provide considerable support for the propositions as Chapters 5 to 9 describe. However, the results are situation specific and so provide only limited benefits to the global construction management community.

Construction management strategies are rarely considered per se. Instead practitioners tend to recognise various ways of undertaking construction actions but even these are predominantly viewed as outside of their companies' control. Nevertheless there are large potential benefits in adopting construction management strategies which enable companies to focus on their core competencies. In fact companies face many choices and can if they chose operate using only a single strategy or adopt a small range of strategies which fit the demands of a selected market. Some companies are already doing this but since there has been no tool available that enable construction management strategies to be compared, most construction companies do no more than simply follow market trends regardless of their capability to execute projects through different strategies. The IDIs defined in Chapter 4 are designed to help companies compare strategies on a case by case basis and so help them identify the optimal strategy for any particular project.

11.3 The theory of construction management and research

Construction management research tends to result from individual ideas and specific situations which occur on individual projects or within single companies. As a consequence the results are fragmented and so lack wide relevance.

This is evident in the following discussion of research which helps in understanding the concepts of agreed objectives, inherent difficulty and construction management strategies.

Agreed objectives

In general construction management researchers accept the view of agreed objectives adopted by the practitioners who provide data for their research. This is sensible from many points of view, not least because it ensures research results are directly relevant to practice. It does however mean the application of research is limited to projects and companies which are very similar to those involved in any specific research. The fundamental cause of this unsatisfactory situation is that it is extremely difficult to measure construction performance in a manner which enables the results to be understood and used generally. In addition, since researchers tend to concentrate on current best practice and work with leading companies, they tend to be inhibited from rationalising their results to envisage theoretical strategies which go beyond current best practice.

The performance of organizations is normally measured in terms of efficiency which is usually defined as the ratio of the value of outputs to the cost of production. In construction this straightforward concept is virtually impossible to apply with any useful degree of consistency. There are many practical causes of this situation.

The value of constructed facilities is difficult to establish, often changes suddenly and dramatically as financial and economic circumstances change, and depends on individual opinions and points of view.

Cost is often regarded as an easier concept to measure and many practitioners act on the assumption that cost can be established reliably. However, in practice cost is influenced by many of the factors which make value so difficult to establish. This is demonstrated by the fact that construction costs often become the subject of disputes on individual construction projects. It is further illustrated by the complex rules which are found to be necessary in cost reimbursable contracts where the customer pays the construction company's legitimately incurred costs.

These practical problems mean measuring efficiency in terms of the ratio of the value of outputs to the cost of production is unreliable. As a consequence it does not provide a robust basis for testing theories about construction performance.

It is common in guides to construction management to define performance in terms of cost, time and quality. The problems with cost are discussed above. Time might appear to be easier to measure accurately. Unfortunately this is far from the case. Establishing the start date for a project is problematic because even straightforward projects are discussed by the interested organizations for days, months or even years before any firm decisions are made. Then the decision to undertake a specific construction project is often reviewed, altered or changed dramatically perhaps in response to new opportunities, regulatory requirements or changed economic or commercial circumstances. The end date is usually easier to determine but even that is not totally straightforward. A common end date is when the completed facility is accepted by the customer. However, sometimes customers insist on moving into a building before it is complete and accept that some production and commissioning activities will continue. It is not uncommon for facilities to include defects when they are handed over to the customer. The defects are put right during the next few months. This raises the question of whether the completion date is when the facility is handed over or when all the defects are corrected. It is also the case that projects can be completed very quickly or very slowly because this suits the customer or the construction companies undertaking the work. So time is not a reliable guide to performance.

The final common criterion for measuring performance is quality. To a large extent this is a subjective measure influenced by similar factors to those which make value difficult to establish reliably. It is further complicated by the way construction quality has become a confused concept.

The specified materials, components and workmanship define the standard of constructed facilities. However, this is often confused with the quality with which they are manufactured, assembled and commissioned. Both the standard and the quality are important but they are different concepts. The standard of a constructed facility is a matter of choice by the customer in discussion with the construction teams employed to produce it. Quality should not be a matter of choice. All constructed facilities should be produced to high quality. In practice the two distinct concepts of standards and quality are muddled and both tend to be subsumed in the one term quality.

Other measures are used to an increasing extent as pressures build up to improve various aspects of construction. In this way safety on construction sites, protecting the environment, contributing to the local economy, low operating costs, architectural excellence, training and construction innovation are often adopted as objectives. It is also argued even if all these factors could be measured rigorously, they would provide a limited basis for construction management decisions. This is because arguably the most important objective for construction is new facilities should function effectively for the customer. Hospitals should promote health, universities should encourage research and learning, motorways should ensure the safe and speedy flow of traffic, and so on.

For the immediate future, IDIs provide researchers with a means of relating established measures of performance, including time and cost, to inherent difficulty. For instance, increased costs may be linked to team performance variability, varying numbers of teams and the quality of relationships between them. Empirical investigations are needed to establish whether such links exist and what exactly the relationships in them look like. Such research will test and fine-tune the construction management theory presented in this book.

International comparisons

One area of construction management research where rigorous attempts to measure performance have been attempted is in comparisons of national construction industries. This research tends to be expensive and time consuming but produces results which influence practice as well as future research. An important example is provided by the results in Table 9.1.

The results in Table 9.1 used data produced by international economic bodies to establish purchasing power parities which included construction as one sector of national economies. The data used related to six building projects and six infrastructure projects in each country. Detailed, standardised descriptions of the projects were priced at current market prices by experienced practitioners in each country. The results were adjusted to purchasing power parity to remove the distortions introduced by market exchange rates and then expressed as an index number with United Kingdom costs used as a datum of 100. The projects were identical across all the countries except where specific details were adjusted to fairly represent normal construction in individual countries. In other words every effort was made to ensure the results measured the cost performance of the various national construction industries.

The results influenced subsequent thinking across construction management practice and research. The research has been repeated by different research bodies on the basis of slightly different combinations of countries. These subsequent studies broadly confirm the two biggest surprises at the time the original research was published, namely, the poor performance of the United States and the realisation that Japan has the most cost efficient construction. Many practitioners believe the United States can match Japan's performance but this conclusion results from comparing the United States' basic, standardised buildings with Japan's higher quality and individually designed buildings and ignoring differences in general price levels. Comparing like with like produces the comparative performance shown in Table 9.1.

Research of this quality is very expensive. The example shown in Table 9.1 required a team drawn from several organizations based in different countries. However, the approach could not provide a robust test of the theory of construction management. No provision is made to match the inherent difficulty faced by the sets of construction projects or ensure the project teams had identical agreed objectives. No attempt was made to ensure the practitioners pricing the descriptions of new facilities assumed any particular construction management strategy. The research was not based on actual construction projects with outcomes which could be reviewed; it was based on current market prices for the elements of constructed facilities.

The IDIs described in Chapter 4 would enable international comparisons to be robust by avoiding traps created by different market and industry structures. IDIs take account of the number of teams, the quality of relationships, performance variability and external interference regardless of local circumstances. Thus, by concentrating on inherent difficulty IDIs would for the first time allow like-for-like international comparative studies. Such comparisons could provide the basis for a robust world performance league table.

It can be concluded that construction practice and research recognises the need for robust measures of performance. However, even the most careful research into a subject as complex as construction performance provokes controversy. The fundamental problem is the absence of a robust and consistent basis which would enable the results to be related and applied widely. A solution to this problem is an important part of the research proposals made later in this chapter.

Inherent difficulty

The inherent difficulty of construction projects has not been well researched. Chapter 4 provides a classification scheme which was essential to allow the theory of construction management to be described. It defines the Inherent Difficulty Indicators, listed in the box on page 266. They provide consistent measures of the basic factors which shape the construction management task on individual projects.

- Established Relationships Indicator measures the incidence of established relationships.
- Relationship Fluctuation Indicator measures the fluctuations in the number of teams and established relationships.
- Relationship Quality Indicator measures the quality of relationships.
- Relationship Configuration Indicator measures the pattern of interactions.
- Performance Variability Indicator measures variations in team performance.
- External Interference Indicator measures the impacts of external factors on performance.

The classification scheme for inherent difficulty includes one other element which is the size of construction projects. The categories identified by the scheme described in Chapter 4 are listed in the box below.

- Minor project involves about 100 construction team days.
- Small project involves about 1,000 construction team days.
- Normal project involves about 5,000 construction team days.
- Large project involves about 10,000 construction team days.
- Mega project involves about 25,000 construction team days.

The proposed classification scheme for the inherent difficulty faced by managers will enable practitioners to recognise projects which pose a broadly similar challenge. It allows them to look at records of the performance levels achieved on similar projects and take this into account in deciding their agreed objectives and selecting a construction management strategy.

Researchers can use the classification scheme to link separate research studies and draw broader and more soundly based conclusions about construction management. They can compare the performance achieved in different parts of national construction industries. It enables them to work across national boundaries with greater confidence.

It can therefore be claimed the Inherent Difficulty Indicators together with the classification of project size set out in Chapter 4 provide a useful basis for practice and research to recognise projects which, in construction management terms, are similar. The development of this first step into a more robust basis for future research is an important part of the proposals made later in this chapter.

Construction management strategies

The construction management propositions provide a guide to the direction of change needed to improve efficiency. They give advice to construction managers in any situation about the actions they should consider when they want to improve project or company performance. In other words they map out stages on a journey. The propositions do not tell construction managers where they are on the journey towards the maximum possible levels of efficiency.

Recognising the construction management propositions define a direction of travel rather than identifying fixed points means they do not provide the basis for classifying construction management strategies. This remains a challenge for the future research described in the following sections.

11.4 Testing the theory of construction management

The theory of construction management provides a basis for coherent construction management strategies. However, the 25 propositions which form the theory need to be tested by research because the only rigorous basis for scientific knowledge is tested propositions.

An essential basis for research designed to test the propositions is that the theory's key concepts are classified in a manner which enables the results to be widely applicable. As described earlier in this chapter in essence this requires a map of every possible construction project so sets of the data needed to test the 25 propositions from individual projects can be related to each other.

Establishing a rigorous map of every possible construction project is not a small task. The map needs to be established on sound principles, used by researchers to relate sets of data from a small number of projects, modified as projects are found which prove difficult to fit onto the map, used with sets of data from a larger number of projects, modified and so on. The following description of principles which could shape the initial version of the map provides a starting point for this process.

The basic factors determining the nature of the construction management task posed by construction projects are the overall size of the project and the specific construction technologies used. The categorisation of project size in terms of team-days described in Chapter 4 and earlier in this chapter provides a sensible basis for the first of these factors. The categorisation of construction technologies will initially at least rely on established practical descriptions. It will help in undertaking the research to use a small number of broad categories. These are likely to include for building projects factors which include: low, medium and high-rise, load bearing and framed structures, minimal, normal and high levels of internal services, and basic, normal and high quality. The categories for infrastructure projects are likely to be based on broad descriptions of the primary function of the completed facility. Thus, it will include motorways, railways, power distribution networks, chemical processing plants, water treatment plants, sewers, water distribution systems, communication networks and other similar facilities. These will be supplemented by sub-categories which reflect the main technological options used in the particular type of infrastructure.

The purpose of the map of every possible construction project is to classify sets of research data designed to test the 25 propositions which form the theory of construction management in such a manner that results can be related to each other in a way that supports further research and provides guidance for practice. However, for the map to be useful it requires sets of research data. Therefore, the next section describes the data needed to test the 25 propositions and proposals for producing the map are described later in this chapter.

11.5 Research data

Research data is required which measures each of the key concepts in the theory. Related to the map of every possible construction project, this data for many different projects provides the material needed to test the 25 propositions.

It also provides other research and practical benefits which are described later in this chapter. The required data is described in the following list.

The number of interacting teams.
>   The number of construction teams;
>   The number of construction team days;
>   The number of time intervals with different team constitutions;
>   The duration of time intervals with different team constitutions; and
>   The number of interacting teams in each time interval with different team constitutions.

Quality of relationships.
>   The number of teams with established relationships in each time interval with different team constitutions.
>   The length of established relationships prior to the current project; and
>   The length of interactions between teams with established relationships.

Performance variability.
>   The percentage of actions each construction team completed early or on time.

External interference.
>   The number and duration of delays caused by external interference.

Agreed objectives.
>   The key agreed objectives.

Project efficiency.
>   Performance in terms of the key agreed objectives; and
>   Major failures in terms of established performance or key agreed objectives.

Further research is likely to provide improvements to some elements of this set of data. It provides a starting point but the following developments are likely. Research may identify a more precise measure of the quality of relationships between interacting construction teams than using the length of established relationships prior to the current project. Research may develop a better measure of performance variability which could be based on measuring daily variations in the output of distinct types of teams. For example, the performance variability indicator may be the same for two projects but there could be vast differences in how much individual team performance variability indicators deviate from the project mean. One project may be plagued by far greater deviations from the mean while the other project may be characterised by a much more consistent distribution of individual performance variability indicators. That is extremely likely to impact overall construction efficiency and requires research based on intra-project investigations to establish what impact such deviations have. Future research may show that data on the causes of significant external interference provides benefits both to research and practice. It is likely as the research becomes an established part of construction management, there will be considerable developments in the data describing agreed objectives and project efficiency.

11.6 Research proposals

Given a map of every possible construction project and a number of sets of the required research data, researchers will be able to start a number of serious investigations. These will be unusually effective because the potential to relate individual results will inevitably raise important new questions. Further the research will provide answers focussed on the direct concerns of serious practitioners. This in turn will create new opportunities and identify important issues for further research.

Individual tests

The most direct way of testing the theory of construction management is to find pairs of projects which test one of the propositions which relate directly to projects. In such a pair one project would act on the proposition to be tested and a second identical project would not. Such pairs of projects would need to be selected very carefully to ensure no factors except the subject of the proposition to be tested influence the performance outcomes of the two projects. Many pairs of projects suitable for testing each proposition would be required to ensure the results can be widely accepted. An important source of such pairs of similar projects is major customers. Some of these undertake many similar construction projects every year and so provide an important source of research data.

Thorough testing of the theory would also require pairs of projects to test every practical combination of the propositions. This whole enterprise would be difficult and time consuming. It would provide robust results and for this reason some researchers will want to test specific propositions in this manner. However, providing a more complete test of the theory requires a more efficient approach capable of producing results within a reasonable timescale and budget.

Simulation

A more efficient approach to testing the propositions combines data from live construction projects with simulations of construction projects.

Simulation can produce a comprehensive map of every possible construction project. The great benefit of such a map is it will allow live projects to be related to each other and the effects of different levels of inherent difficulty on the same project to be explored. For instance, a live project involves a particular number of teams but the effects of increasing or decreasing the number of teams can be examined through simulation. In the same way all the factors which influence the choice of construction management strategies can be studied.

The first step in providing this comprehensive map is to establish a theoretically robust method of simulating the outputs of construction projects which take account of their agreed objectives, inherent difficulty and construction management strategies. Having developed a robust simulation scheme, the next step in the research is to use data from live projects to set up and test the simulation scheme. This is a process of using input data from a live project, running the simulation and comparing the outputs with the actual project outputs. Then the simulation can be adjusted until the simulated and actual outputs coincide. Ideally many live projects will be used to tune the simulation scheme.

A number of approaches could be used for simulation purposes. Case Based Reasoning (CBR), agent-based modelling and social networks are just some techniques that could be applied. Also chaos theory could provide the basis for insights into the dynamic nature of IDIs. However, it is likely that the most effective method of simulating the outputs of construction projects will result from the above techniques being incorporated in BIM. This is because the theory of construction management will encourage designers and construction managers to demand efficient ways of evaluating tentative construction management strategies at the earliest possible stages of projects. Data from previous similar projects will enable BIM to evaluate preliminary designs in terms of the probable IDIs. This in effect provides an early risk assessment and if the results suggest the project is likely to fail in some respects, BIM will allow alternative designs, objectives and strategies to be considered sufficiently early to allow changes to be made with minimal risk of wasting time or resources. Using an analogy with CAD libraries, BIM could include databanks of IDI-ranked alternative designs corresponding to a particular local construction industry, and information on past projects, including past team performance, quality of relationships and other IDI information requirements. These need to be developed but BIM software is sufficiently adaptable and interoperable to integrate a growing pool of databanks.

Virtual map of construction project performance

Having devised an effective simulation scheme, the next step is to define a complete set of construction projects in terms of the essential input data. The set will include sufficient projects to cover the complete range of objectives, inherent difficulty indicators, project size categories and combinations of construction management propositions. In other words it provides a virtual map of every theoretically possible construction project.

It is unlikely anyone would ever feel the need to simulate enough projects to produce the complete map which would be multi-dimensional and extremely large. So it remains a virtual map implicit in the scheme which is capable of simulating the performance of every possible construction project.

Testing the propositions which relate to projects

Research can then use the simulation scheme to test the theory that every application of the propositions individually or in combination delivers improvements in performance for every category of construction project. Researchers can test individual construction management propositions and interesting combinations of them on a range of projects to explore ideas which they find interesting or which are of interest to some part of the construction industry.

Research which provides data from live projects can be plotted onto the theoretical map. This will allow individual research to be related to a wide body of research results. It will at the same time further validate the simulation scheme and test the construction management propositions.

It is of course possible that some research results will show a proposition is not supported. In this situation each step in the research needs to be checked to ensure the results are not produced from mistakes in classification, data collection or input. The implementation of the propositions should be questioned to ensure they have been put into practice effectively. Where the research appears to be robust and does not support the proposition or propositions, this important result should be published. In parallel with publishing the initial results, an attempt should be made to re-test the same set of propositions. If repeated tests show a proposition is not supported, the theory will need to be changed.

Testing the propositions which relate to companies

Testing the propositions which relate to company behaviour is more complicated. Effective research depends on the virtual map of construction project
performance having been developed into a robust form. Given the robust map
exists, the research begins by identifying companies acting on the basis of one
or more of the propositions. Ideally a company will act on the basis of just one of
the propositions which relate to company behaviour but many companies will
make more complex changes to their policy. In either case the company will
undertake a series of projects some of which occur prior to the change of policy
and some occur after the proposition or propositions are put into effect. The
project outcomes are compared to find out if the predicted benefits have been
achieved.

Robust results will require the research to cover a number of companies
adopting a variety of policy changes which influence a number of projects. The
resulting data will be complex and understanding it depends on being able to
plot it onto the virtual map of construction project performance. Where the
research shows project performance improved relative to the norms provided
by the virtual map after the propositions were implemented, the theory is
supported.

A more difficult situation arises where the research does not support
the propositions. The practitioners involved in the company have a problem
and no doubt will review their policy and its application. There are many
possible reasons for a policy based on the construction management propositions to fail. It may have been introduced badly, not explained to key
individuals, be inappropriate to the company's objectives, be undermined by
individuals who resent change, have been applied to projects which provided
conditions alien to the proposition, or the proposition may be wrong. The
researchers need to adopt the same rigorous approach described above in
respect of propositions relating to projects. They need to check their
research carefully and attempt new tests but they must recognise it is always
be possible for a proposition to be contradicted. When this occurs, the
research which made this negative discovery should be published and the
theory changed.

Research questions

The map of every possible construction project and the required data sets will
help answer many important research questions. Those listed in this section are
examples selected to deal with each of the key concepts used in the theory of
construction management.

Research Questions about the Number of Interacting Teams

- What is the optimum number of teams for projects of each size category
  and distinct construction technology in terms of construction efficiency?
- What is the optimum configuration of teams in terms of the number of
  time intervals, the duration of time intervals, and the number of interacting teams in each time interval for projects of each size category and
  distinct construction technology?

An effective way of answering these research questions is for the construction management research community to study projects worldwide, ideally concentrating on counties with the highest levels of construction efficiency. The first question requires data describing the number of construction teams and the achieved level of efficiency to be collected about sufficient projects to include multiple examples of each size category and distinct construction technology.

The second question requires, for the same range of projects, data describing the number of time intervals, the duration of time intervals, and the number of interacting teams in each time interval. This data should relate to all the distinct actions which form the complete project. This is emphasised because there is some tendency for construction management research to deal with production and give little attention to the other construction actions.

Answers to these questions will guide practice as they make decisions about project organizations. They should check their proposed strategy results in about the same number of teams, a similar pattern of time intervals and interacting teams as on the most efficient projects which are similar to their own. Where they find significant differences, they should reconsider their approach.

Research Questions about the Quality of Relationships

- What is the influence of established relationships on construction efficiency?
- Which relationships in projects of each size category and distinct construction technology benefit most from established relationships in terms of construction efficiency?

The theory of construction management assumes established relationships improve construction efficiency. The proposed research data further assumes established relationships improve directly with the length of time teams have worked together on previous projects. The rational for this second assumption is essentially practical. Teams that have worked together on projects where their relationships were ineffective are less likely to be required to work together again. Hence, it is likely that teams that have worked together on several previous projects have well-established relationships that have been replicated because everyone involved is keen to continue the good work. Answers to these questions will allow these assumptions to be tested and if necessary changes made to the research data used to indicate the existence of effective established relationships.

The answers will more directly help practice understand the benefits of established relationships on projects of different sizes using distinct construction technologies. Perhaps more significantly, knowing which relationships benefit most from established relationships will guide their choice of construction teams.

Research Questions about Performance Variability

- What are the most significant causes of teams having distinct levels of performance variability?
- What are the effects of distinct patterns of performance variability on construction efficiency?

The first of these questions goes beyond the research data but is guided by it. As distinct patterns of performance variability are identified on individual projects, researchers will attempt to identify reasons for the differences. The reasons may include differences in the use of well-established and innovative technologies, different approaches to education and training, different employment arrangements, different levels and types of support from companies and project organizations, and many other possible causes.

The answers will help practice recognise the importance of established relationships on projects similar to those they undertake. They will also suggest ways of fostering and supporting established relationships.

Research Questions about External Interference

- What are the distinct categories of events which cause external interference in terms of their nature, duration and magnitude?
- What is the impact on construction efficiency of each distinct category of events which cause external interference?

External interference is difficult to research because specific occurrences are largely unpredictable. This usually means researchers become involved only after some major external interference has occurred on a project. Research at this time is often particularly difficult because construction teams struggling to overcome the effects of external interference tend to be unwilling to spare time to discuss their immediate situation with researchers. This means the information needed to answer these questions is most likely to result by chance because researchers are already involved on a project which suffers external interference and they recognise an opportunity to collect valuable data.

Practice is likely to become more willing to cooperate with research into external interference if they can be given useful results from earlier research. Such results are likely to help practitioners identify major risks on projects similar to their own. Knowing the potential nature, duration and magnitude of external interference and its impact on a project's overall efficiency helps decisions about precautions which are worth adopting and those unlikely to provide net benefits.

Research Questions about Agreed Objectives

- What distinct categories of agreed objectives exist to guide construction project teams?
- What influence does each category of agreed objectives have on the optimum number of teams, the influence of established relationships, and the optimum configuration of teams?

It is likely to be difficult for researchers to identify the key agreed objectives on individual projects. Research shows the effective objectives which guide project organizations often change throughout the course of projects and it is not uncommon for individual teams within project organizations to have different, even conflicting objectives. It is however becoming more common, particularly in projects where the strategy recognises the benefits of fostering cooperation between the companies involved, for the objectives to be explicitly agreed. The most effective research into agreed objectives will come from such projects.

Practice will benefit from seeing the benefits of formally agreed objectives on construction efficiency and from guidance on the implications for their strategic options of specific categories of agreed objectives.

Research into inherent difficulty indicators

Another benefit of the map of every possible construction project and the required data sets is they will allow the IDIs described in Chapter 4 to be tested and improved. The following descriptions suggest key research issues relating to each of the IDIs.

Equation 4.13 provides the indicator for established relationships. It straight-forwardly measures the incidence of established relationships among the construction teams involved in a project. Research may identify differences in the benefits delivered by established relationships on projects of different sizes or those using different technologies. Research may show established relationships at the various stages of construction projects have different influences on construction efficiency. Research may also suggest established relationships are most effective when the IDIs reach particular values.

Equation 4.15 provides the indicator for relationship fluctuation. It measures changes in the number of established relationships over separate time intervals throughout a project. Research will identify the impact on construction efficiency and in turn suggest how the construction management task is affected by having to deal with stable project organizations where established relationships exist uniformly across the whole project duration compared with those which change frequently. Fluctuation across time intervals can be important even when the corresponding IDI has the same value in different projects. Research may show fluctuations at particular times in a project have significant impacts on overall performance.

Equation 4.17 provides the indicator for relationship quality. It measures the quality of relationships on a project in terms of the length of time teams have worked together on previous projects, taking account of the relative importance of interactions. Research may identify better indicators of relationship quality than simply using the length of time teams have worked together on previous projects. It will also provide ever more detailed guidance on the relative importance of the main relationships on projects of various sizes and distinct construction technologies. These results are likely to have the important benefit of focussing practice on the importance of fostering high quality relationships.

Equation 4.18 provides the indicator for relationship configuration. It measures the patterns of relationships between construction teams involved in a project. It comprises three elements which represent the number of time intervals with different patterns of interactions, the influence of the length of time intervals, and the influence of the number of interacting teams. Research is needed to establish the correct weighting for each of these elements as it is unlikely the approach used in the various examples earlier in the book of assuming the elements are equally important is correct for all projects. The results will help practice to understand the impacts on construction efficiency of changes to the configuration of project organizations and lead to greater emphasis being given to maintaining efficient patterns. Equally it may emerge there are several efficient patterns and research may be able to identify which of these suit which particular circumstances.

Equation 4.20 provides the indicator for performance variability. It straight forwardly measures the performance variability track record of the construction teams involved in a project. Various research issues concerning the way this important factor is measured are described earlier in this chapter. Beyond that essentially practical issue, this IDI will focus practitioners' attention on the important effects of performance variability on construction efficiency. This in turn will create demands for research to identify effective ways of reducing performance variability.

Equation 4.21 provides the indicator for external interference. It straightfor-wardly measures the external interference experienced by similar projects. As discussed earlier in this chapter, this IDI will provide useful information for practitioners undertaking risk analysis on their projects. Research is likely to concentrate on the causes of major impacts on projects caused by external interference and strategies which reduce its incidence and impacts.

The IDIs are designed to provide an early indication of the inherent difficulty of any given project to help practitioners select appropriate construction management strategies. Research will establish the usefulness of this prelimi-nary list of IDIs and over time lead to improvements and changes which make the indicators ever more effective.

11.7 A basis for future practice and research

The research described in this chapter would give the construction manage-ment research community a robust basis for designing and undertaking future research. Most importantly it will provide construction management practice with a robust basis for selecting the most appropriate construction manage-ment strategy for any given project or company. Developments in BIM should ensure construction managers have the tools and data needed to evaluate alternative strategies sufficiently early to enable them to be fully involved in strategic decisions. This particularly refers to information needed to calculate IDIs, which needs to be available before these strategic decisions are made. These benefits are essential elements in raising the level of construction efficiency so it becomes a truly modern industry equal to any other major sector of the global economy.

Exercise

Consider starting research into hotel projects from all over the world. Out of a total of 50 projects 12 are in Europe, 15 in South East Asia, 9 in South America and 14 in North America. This will clearly be a comparative study and you may well produce a virtual map of world hotel project performance. Take into account the construction management theory and IDIs in Chap-ter 4, think about the following questions and formulate a coherent research strategy that would produce the desired virtual map:

1. What kind of project information would you require to be able to calculate the IDIs?

2. What kind of project information would you require in order to establish what the selected construction management strategies were on individual projects?
3. What data collection methods would you use to obtain the necessary project information and how would you plan data collection?
4. Could you simulate performance on these projects and if so, what techniques would you use for simulation?
5. What benefits could practice gain from the results of the research?

Further Reading

The following publications are the source of ideas used in this chapter and provide further information for readers.

Bennett, J. and Ormerod, R.N. (1984) Simulation applied to construction projects. *Construction Management and Economics*, 2(3), 225–63. This classic paper illustrates the care and detail needed to simulate live construction projects. Modern computing and the data now available from the global construction management research community now make the task considerably less daunting.

Pickrell, S., Garnett, N. and Baldwin, J. (1997) *Measuring Up: A Practical Guide to Benchmarking in Construction*. Construction Research Communications, London. This guide is based on good research into the use of benchmarking in construction and provides a straightforward guide to best practice.

Prietula, M., Carley, K. and Gasser, L. (Eds.) (1998) *Simulating Organizations: Computational Models of Institutions and Groups*. AAAI Press. This book is a collection of fascinating articles about computational modelling of organizations from a variety of perspectives. The articles provide a really good starting point for researchers interested in better understanding complex changes that organizations are undergoing on a daily basis, many of which can only be studied through computational modelling.

Appendix

Theory of construction management propositions

1. Construction management aims to reduce the number of construction teams involved in a construction project.
2. Construction management aims to improve the quality of relationships between the construction teams involved in a construction project.
3. Construction management aims to reduce the performance variability of the construction teams involved in a construction project.
4. Construction management aims to reduce the external interference experienced by the construction teams involved in a construction project.
5. Construction management aims to select construction teams competent in the technologies required by the project in which they are involved.
6. Construction management aims to ensure construction teams accept the agreed objectives of the project in which they are involved.
7. Construction management aims to ensure construction teams are motivated to achieve the agreed objectives of the project in which they are involved.
8. Construction management aims to foster accurate communication between the construction teams involved in a project organization.
9. Construction management aims to minimise the effort required to achieve accurate communication between the construction teams involved in a project organization.
10. Construction management aims to minimise the length and intensity of negotiations needed to agree the transaction which bring construction teams into a project organization.
11. Construction management aims to ensure construction teams regard the transaction which brought them into a project organization as advantageous to themselves.
12. Construction management aims to minimise the resources construction teams devote to improving the terms of the transaction which brought them into a project organization.
13. Construction management aims to satisfy the requirements of local or specialised construction markets in ways which ensure their company's long-term survival.
14. Construction management aims to develop construction teams with well-developed skills and knowledge which match the requirements of construction projects.
15. Construction management aims to develop construction teams integrated by established relationships.

Construction Management Strategies: A Theory of Construction Management,
First Edition. Milan Radosavljevic and John Bennett.
© 2012 John Wiley & Sons, Ltd. Published 2012 by John Wiley & Sons, Ltd.

16. Construction management aims to improve the quality of support provided to construction teams by the construction companies of which they form part.

17. Construction management aims to foster innovations which match the requirements of construction projects.

18. Construction management aims to ensure their organization uses effective information systems.

19. Construction management aims to establish values which run consistently through every aspect of their organization's work.

20. Construction management aims to ensure all parts of their organization act in ways which are intended and authorised.

21. Construction management aims to ensure their organization's communications are effective and result in common understandings.

22. Construction management aims to ensure their organization's transactions are agreed with the minimum of effort, accepted as fair by all the parties involved, foster established relationships, and are acted on in the spirit in which they were agreed.

23. Construction management aims to ensure their organization forms established relationships with other organizations.

24. Construction management aims to ensure their construction organization collects, reviews and acts on feedback about the effects of the organization's actions on its objectives.

25. Construction management aims to ensure their construction organization collects, reviews and acts on feedback about the effects of established norms and procedures.

Glossary

Most terms used in the book rely on normal definitions found in good dictionaries. The following terms have a more precise definition in the theory of construction management.

Agreed objectives means the set of aims which motivate organizations responsible for undertaking construction.

Boundary relationship is a relationship between organizations in which their behaviour is guided by their perception they are parts of different organizations.

Brief is a statement describing a new facility required by a customer and identifying constraints which must be observed in undertaking the necessary construction actions.

Built environment is all currently existing constructed facilities.

Commissioning is the actions which turn the product of production into a fully tested and properly functioning new facility and ensure the organization which will take over and run the new facility is trained to use and operate it.

Communication is transfer information of between organizations.

Complexity is a measure of the number of interacting construction teams involved in a construction project, the quality of the relationships between them and their performance variability.

Composing is the process of developing internal relationships.

Construction is all human actions intended to produce and/or alter facilities.

Construction company is a permanent organization which undertakes construction actions and so supports one or more construction teams.

Construction company division is a distinct organizational part of a construction company.

Construction industry is the totality of construction companies currently operating.

Construction management is taking responsibility for the performance of a construction organization.

Construction management strategy is a coordinated set of decisions which guide a construction project organization.

Construction organization is a formal group of individuals who undertake construction actions. Thus, the term includes construction teams, companies, company divisions, and project and programme organizations.

Construction programme is a series of construction projects which have sufficient characteristics in common to justify treating them as an integrated whole.

Construction programme organization is the set of construction teams responsible for undertaking a construction programme.

Construction Management Strategies: A Theory of Construction Management,
First Edition. Milan Radosavljevic and John Bennett.
© 2012 John Wiley & Sons, Ltd. Published 2012 by John Wiley & Sons, Ltd.

Construction project is a set of the seven essential actions needed to produce a new facility which have a start and end date. They are producing a brief, design and plan, and procurement, manufacturing, production and commissioning.

Construction project organization is the set of construction teams responsible for undertaking a construction project.

Construction team is a formal group of individuals who work together on a permanent basis to undertake specialist construction and the essential machines and equipment the team uses.

Customer is an individual or organization external to the construction industry which initiates construction actions.

Design is a statement describing the product of construction actions which satisfy the requirements of the brief for a new facility.

Efficiency is a measure inversely related to the waste caused by complexity and external interference which prevent organizations achieving their agreed objectives.

Established relationship is a relationship between organizations which have worked together efficiently over a significant period of time.

External interference is a measure of the impact of factors external to a construction project.

Facility is a permanent, fixed product constructed by human organizations.

Feedback is information about the effects of an organization's own actions.

Folding is the process of forming boundary relationships.

Information is remembered and recorded useful knowledge.

Inherent difficulty is a measure of the complexity and external interference experienced by a construction project organization using traditional construction practice.

Interaction is communication or transaction between organizations.

Internal relationship is a relationship between construction teams in which their behaviour is guided by their perception they are parts of a joint organization.

Management is taking responsibility for the performance of an organization.

Manufacturing is the actions needed to make and provide the materials and components needed for production.

Network is a set of interacting organizations and the relationships between them.

Owner is an organization which is the legal possessor of a facility.

Performance variability is a measure of the range of performance achieved by a construction team.

Plan is a statement identifying the types of organization needed to produce a new facility and describing the constraints and targets which influence the way these organizations need to work.

Procurement is the actions which ensure construction teams with the required skills, equipment and support are selected and employed and arrangements are in place so the teams are supplied with the necessary materials and components and are motivated to undertake the construction needed to produce a new facility.

Production is the actions which convert materials and components into a new facility.

Quality of relationships is a measure inversely related to the time and resources devoted to interactions by construction teams.

Relationship is a linked series of interactions.

Traditional construction is the locally established actions of construction project organizations.

Transaction is exchange of things of value between organizations.

Values are the positions allocated to things in an order of preferences.

Index

accident, 211, 233
acquisition, 142–3, 242
action, 54–9, 69, 70, 240, 241
  authorised, 64, 93, 221, 239,
    244, 254
  coordinated, 65, 148
  specialised, 58, 162, 253–4
  standardised, 186
advertising, 190, 242
advice,
  expert, 163, 232,233
  independent, 231–2
  strategic, 231
  technical, 24
agent based modelling, 269
agreed objectives, 89–90, 95–6, 132,
    133, 149, 150, 166, 168, 177, 179–80,
    194, 217, 220, 234–6, 237, 247, 249,
    253, 258, 259–61, 263–4, 273–4
  defined, 95
  example, 259–60
  ownership, 260–61
agreement, 143, 217
  formal, 189
  win-win, 179
airport, 31, 46
approval, 9, 160
architect, 22–3, 24, 60, 115, 121, 132–3,
    138, 140, 159
  control, 128–30
  coordination, 127–30
  design, 48–9, 112
  detail design, 127–30, 159
architecture, 37–9
  American, 49
  British, 22–3, 49
  consistent, 147–8
  high quality, 23, 159, 264
  Japanese, 26–7, 49
  outstanding, 23, 147–8, 159
  signature, 252

Arup Associates, 158
attitudes, established, 188
automation, 202–03, 216, *see also*
    production, automated
Axelrod, R., 72

Barcelona, 50
Beauvais Cathedral, 113
Beijing, 3, 40, 42
benchmarking, 261
benchmarks, 186, 261
benefit, 87, 182, 187, 233, 234, 237
  distribution, 184
Bennett, J., 29, 226, 227
Bennett, J. and Jayes, S., 200
Bennett, J. and Ormerod, R.N., 276
Bennett, J. and Peace, S., 200
Bennett, J., Pothecary, E. and Robinson,
    G., 154
Bertelsen, S., and Sack, R. 29
best buy, 230
bidding, 22, 23, 57, 65, 118, 120, 129, 139,
    144, 163, 167, 177, 238
  low, 122
bills of quantities, 22–3, 129–30
Bluewater shopping centre, 159
boundary relationship, 67–8, 78–80, 90,
    112, 118–19, 122–3, 142
  defined, 68
  formal, 118
  internal, 80, 142, 242
Bovis, 158
Brandon, P.S. and Lombardi, P., 52
Brasília, 42
Bresnen, M., 200
brief, 54. 77, 80, 92, 232, 234, 244, 248,
    see also construction management,
    design build, developed traditional
    construction, partnering, total
    construction service
  defined, 54

Printed and bound by CPI Group (UK) Ltd, Croydon, CR0 4YY

27/10/2024

14580291-0001